18 —

90/43

THE SEMI-CENTENNIAL ANNIVERSARY

ANNIVERSARY

OF THE

NATIONAL ACADEMY OF SCIENCES

This is a volume in the Arno Press collection

THREE CENTURIES
OF
SCIENCE IN AMERICA

Advisory Editor
I. Bernard Cohen

Editorial Board
Anderson Hunter Dupree
Donald H. Fleming
Brooke Hindle

See last pages of this volume for a complete list of titles

THE SEMI-CENTENNIAL ANNIVERSARY

ANNIVERSARY

OF THE

NATIONAL ACADEMY OF SCIENCES

1863-1913

Vol. II

ARNO PRESS

A New York Times Company
New York • 1980

Editorial Supervision: Steve Bedney

―――――

Reprint Edition 1980 by Arno Press Inc.
Reprinted from a copy in the University of Illinois Library
THREE CENTURIES OF SCIENCE IN AMERICA
ISBN for complete set: 0-405-12525-9
See last pages of this volume for titles.
Manufactured in the United States of America

―――――

Library of Congress Cataloging in Publication Data

National Academy of Sciences, Washington, D.C.
    The semi-centennial anniversary of the National
Academy of Sciences, 1863-1913.

    (Three centuries of science in America)
    Reprint of the 1913 ed. published by the Academy,
Washington.
    "Vol. 2."
    1. National Academy of Sciences, Washington, D. C.--
History--Addresses, essays, lectures. I. Title.
II. Series.
Q11.N286 1980a   353.008'55   79-24941

THE SEMI-CENTENNIAL
ANNIVERSARY

OF THE

NATIONAL ACADEMY
OF SCIENCES

1863–1913

THE SEMI-CENTENNIAL ANNIVERSARY

OF THE

NATIONAL ACADEMY OF SCIENCES

1863-1913

WASHINGTON

1913

The Lord Baltimore Press
BALTIMORE, MD., U. S. A.

PREFACE

IT seemed desirable to the committee in charge of the arrangements for the Semi-Centennial Meeting of the National Academy of Sciences, held at the Smithsonian Institution in Washington on April 22, 23, and 24, 1913, to include in the plans some provision for a permanent record of its proceedings.

Accordingly, a contract was entered into with a local reporting bureau to make duplicate stenographic reports of the scientific papers, the presentation addresses, and of the after-dinner speaking at the formal dinner of the members and their guests on the closing day of the meeting.

This record, after careful comparison of the two sets of stenographic notes and, in the case of certain of the scientific papers, with the author's manuscript, was thought to be a true and complete record of the proceedings of the Semi-Centennial Meeting, and in this form, without editing of any kind, it was presented to the Council of the Academy on May 21, 1913.

The Council directed that this record be printed.

In the closing pages of the volume there has been added a record of the attendance in the form of a reproduction of the signatures of all the members and guests registered at the meeting.

The Secretary offers this brief preface partly in explanation of the origin of this volume, and partly in certification of the correctness of the record in so far as it has proved possible to establish it.

ARTHUR L. DAY,
Home Secretary.

Washington, November 18, 1913.

TABLE OF CONTENTS

1863  1913

Semi-Centennial Anniversary

of the

National Academy of Sciences

Smithsonian Institution

Washington

PROGRAM.

TUESDAY, APRIL 22

9.30 A. M. BUSINESS MEETING OF THE ACADEMY IN THE NATIONAL MUSEUM.*
MEMBERS OF THE ACADEMY ONLY.

11.00 A. M. OPENING SESSION, NATIONAL MUSEUM.*
MEMBERS OF THE ACADEMY AND GUESTS.

WELCOME BY THE PRESIDENT OF THE ACADEMY.

ADDRESSES:
"THE RELATION OF SCIENCE TO HIGHER EDUCATION IN AMERICA."
PRESIDENT ARTHUR T. HADLEY
OF YALE.

"INTERNATIONAL COOPERATION IN RESEARCH."
DR. ARTHUR SCHUSTER
SECRETARY OF THE ROYAL SOCIETY OF LONDON.

1.30 P. M. LUNCHEON IN THE GRILL ROOM OF THE HOTEL RALEIGH.
MEMBERS OF THE ACADEMY AND GUESTS. ADMISSION BY CARD.

3.00 P. M. AFTERNOON SESSION. NATIONAL MUSEUM.*
MEMBERS OF THE ACADEMY AND GUESTS.

ADDRESS:
"THE EARTH AND SUN AS MAGNETS."
DR. GEORGE E. HALE,
DIRECTOR OF THE MOUNT WILSON SOLAR OBSERVATORY.

9.00 P. M. RECEPTION BY THE REGENTS AND SECRETARY OF THE SMITHSONIAN
INSTITUTION TO THE MEMBERS OF THE ACADEMY AND INVITED
GUESTS AT THE NATIONAL MUSEUM.
PLEASE PRESENT THE CARD OF INVITATION.*

*NORTH ENTRANCE, 10TH AND B STREETS N. W.

WEDNESDAY, APRIL 23

10.30 A. M. MORNING SESSION. NATIONAL MUSEUM.*
MEMBERS OF THE ACADEMY AND GUESTS.

ADDRESSES:
"ON THE MATERIAL BASIS OF HEREDITY."
DR. THEODOR BOVERI,
UNIVERSITY OF WÜRZBURG.
"THE STRUCTURE OF THE UNIVERSE."
DR. J. C. KAPTEYN,
DIRECTOR OF THE ASTRONOMICAL LABORATORY, UNIVERSITY OF GRONINGEN.

1.00 P. M. LUNCHEON IN THE GRILL ROOM OF THE HOTEL RALEIGH.
MEMBERS OF THE ACADEMY AND GUESTS. ADMISSION BY CARD.

3.30 P. M. RECEPTION AT THE WHITE HOUSE AND PRESENTATION OF MEDALS BY
THE PRESIDENT OF THE UNITED STATES.
MEMBERS OF THE ACADEMY AND GUESTS. ADMISSION BY CARD.

9.00 P. M. RECEPTION BY THE TRUSTEES OF THE CARNEGIE INSTITUTION OF WASH-
INGTON** TO MEMBERS OF THE ACADEMY AND INVITED GUESTS.

THURSDAY, APRIL 24

9.00 A. M. MEETING OF THE COUNCIL AT THE NATIONAL MUSEUM.*

9.30 A. M. BUSINESS MEETING OF THE ACADEMY AT THE NATIONAL MUSEUM.*
MEMBERS OF THE ACADEMY ONLY.

10.00 A. M. DURING THE BUSINESS MEETING OF THE ACADEMY, OPPORTUNITY WILL
BE GIVEN TO GUESTS TO VISIT THE SCIENTIFIC BUREAUS AND
LABORATORIES OF WASHINGTON. AUTOMOBILES WILL BE PROVIDED.
DETAILS WILL BE FOUND AT THE BUREAU OF INFORMATION, NATIONAL MUSEUM.*

12.30 P. M. LUNCHEON IN THE OAK ROOM OF THE HOTEL RALEIGH.
MEMBERS OF THE ACADEMY ONLY. ADMISSION BY CARD.

2.00 P. M. EXCURSION TO THE HOME OF WASHINGTON AT MOUNT VERNON BY
THE U. S. S. "MAYFLOWER."
BY COURTESY OF THE SECRETARY OF THE NAVY. ADMISSION BY CARD.

8.00 P. M. DINNER AT THE NEW WILLARD HOTEL.

* NORTH ENTRANCE, 10TH AND B STREETS N. W.
** CORNER P AND 16TH STREETS N. W.

THE SEMI-CENTENNIAL ANNIVERSARY

OF THE

NATIONAL ACADEMY OF SCIENCES

SMITHSONIAN INSTITUTION, WASHINGTON, D. C.,
Tuesday, April 22, 1913.

MORNING SESSION

The opening session was called to order by the President of the Academy, at 11 : 00 o'clock a. m.

WELCOME BY THE PRESIDENT OF THE ACADEMY

PRESIDENT REMSEN: As presiding officer of the Academy for the time being, it is my duty to express our gratification at the coming together of our guests and to extend to you a welcome.

We have come together especially to take note of the fact that fifty years ago a number of prominent workers in the field of science founded the National Academy of Sciences, receiving a charter from the United States Government. It would be interesting and instructive to call the roll of the founders and learn who they were, but it will suffice to refer to some of the most eminent or most conspicuous among these, or perhaps it would be better to say some of those whose names are most familiar to the present generation.

High up on this honor list are Louis Agassiz, James D. Dana, Wolcott Gibbs, B. A. Gould, Asa Gray, A. Guyot, Joseph Henry, J. Leidy, J. P. Lesley, Benjamin Peirce, R. E. Rogers, W. B. Rogers, L. M. Rutherfurd, Benjamin Silliman, Jeffries Wyman and J. D. Whitney.

Fifty names are included in the Act of Incorporation. Among these are several members of the United States Army and Navy, as for example J. G. Barnard, J. A. Dahlgren, Charles H. Davis, John Rodgers, J. G. Totten, and others holding positions in the United States Military Academy and the United States Naval Observatory and the Naval Academy.

A careful scrutiny of the list of incorporators will show that they can be classified under three heads. The majority were engaged in scientific researches and had reached results of value. They were the leaders among the scientific investigators of that day. Then there were those who had gained distinction by their services as engineers, either in the army or navy; and a third class was composed of heads of the national institutions, such as the United States Naval Observatory, Naval Academy, Military Academy and Coast Survey.

Section 2 of the Act of Incorporation provides that the Academy, " . . . . shall consist of not more than fifty ordinary members, and . . . . shall have power to make its own organization, including its constitution, by-laws, and rules and regulations."

Nothing is said in regard to the qualifications for membership. This is equally true of the constitution and rules, except that Article I, Section 1 of the constitution, requires that," Members must be citizens of the United States." It should, however, be noted that Article IV, Section 4, of the constitution, contains this clause: " . . . . Each nomination shall be accompanied by a list of the principal contributions of the nominee to science." The reference is to nominations for membership, and the inference is clear that the nominee was assumed to have original works to his credit.

Whatever may have been the views of the incorporators, it has gradually come to be held that membership should stand for successful activity in the field of scientific research, the word " scientific " as here used meaning that which pertains to the natural sciences, and-physical sciences, if there is a difference. But our predecessors did not intend to bind themselves to this meaning, as is clearly shown by the election of James Hadley in 1864, G. P. Marsh in 1865, and later of Francis A. Walker and Richmond Mayo-Smith.

As regards engineers who were prominently recognized in the early days of the Academy, the change of attitude that is worthy of notice, is, briefly, this: While one who had accomplished some engineering feat was formerly regarded as worthy of membership by virtue of that fact, now the view appears to prevail—there is no constitutional provision—that only such engineers as have advanced their subject by original contributions should be recognized.

Finally, it is no longer held that the heads of scientific bureaus or departments of the United States Government should necessarily be made members of the Academy, no matter whether they have been actively engaged in scientific research or not. It is evident, therefore, that the field of choice has gradually become narrower.

What was perhaps regarded as the most important part of the Act of Incorporation is contained in Section 3, and reads as follows:

".... the Academy shall, whenever called upon by any Department of the Government, investigate, examine, experiment and report upon any subject of science or art, the actual expense of such investigations, examinations, experiments and reports to be paid from appropriations which may be made for the purpose, but the Academy shall receive no compensation whatever for any services to the Government of the United States."

This clause is still valid. The United States Government may at any time call upon the Academy for investigations, opinions, and advice on any subject of science or art, and this without charge for services.

In order the more clearly to understand the situation that existed in 1863, we should bear in mind two facts: First, there were at that time but few scientific bureaus forming parts of the National Government; and, second, it was a time of war.

Perhaps it would be better to state these facts in the other possible order. Service to the Government was uppermost in men's minds. If they could not help in one way, they could in another. What more natural than this willingness to place their knowledge and skill in scientific matters at the disposal of the Government? This was an act of patriotism, and patriotism was in the air.

While engineers, astronomers and mathematicians were then well represented in the works of those who were serving their country in one capacity or another, it was a difficult matter for those in authority to secure authoritative opinions and advice in other branches of science. There was a gap to be filled. By granting a charter to a group of the leading workers in all branches of science on the terms under consideration, the gap was filled in what appeared to be a most satisfactory manner. After that act, there could no longer be excuse for not seeking scientific advice whenever it was desired or needed.

How did this work? An examination of the records shows that for a number of years after the National Academy was incorporated, the Government frequently called for reports. Six such reports were made in the first year of the existence of the Academy. The subjects were:

" On the protection of the bottoms of iron vessels."
" On the magnetic deviations in iron ships."
" On an alcoholometer."
" On the explosion of a boiler on the United States gunboat *Chenango.*"
" On the use of aluminium bronze for cent coinage."
" On wind and current charts and sailing directions."

In 1865 there were two reports; in 1866, four; in 1867, two, both of which are worthy of special mention. They are:

" On the improvement of Greytown Harbor, Nicaragua," and
" On galvanic action from association of iron and zinc."

In 1868 there were two. In 1870 there was one report, " On the protection of coal mines from explosions by electricity "; another, " On removal of ink from revenue stamps," and a third, " On silk culture in the United States."

In 1875 and 1876 there was only one each. Then in 1878 there were several important reports—six in all—among them one "On proposed changes in the Nautical Almanac," another, "On the use of polarised light for determining values of sugars," another "On the measurement of the velocity of light," and another "On the preservation of the writing of the original Declaration of Independence." A second report on that subject was asked for only two or three years ago, and resulted in the consigning of the Declaration of Independence to a safe, where it would be protected from light.

While there have been important reports on important subjects since 1878, it is undoubtedly true that of late years the Academy has been called upon less frequently than in the early years. At first the officers of the National Government took the matter seriously, and this was to the advantage of the country. But with the multiplication of scientific bureaus supported by the Government, the need of help from the Academy has become less. While it is true that some of the subjects already mentioned, and others not mentioned, could have been reported upon by one or another of the bureaus now in existence, the conditions have changed, as already stated, but even as matters now stand, there is ample room for the kind of activity which was in the minds of the founders. Large questions of a scientific character present themselves from time to time, and it is hard to conceive of a better method of dealing with such questions than that under consideration.

In this connection it should be borne in mind that advice, even good advice, is not always heeded. Indeed, it may happen that it is treated almost contemptuously. This is well illustrated by an actual case which deals with an important governmental problem. Owing to its importance, this case may well be treated of in some little detail.

The Sundry Civil Act, approved May 27, 1908, requests the National Academy of Sciences to consider certain questions relating to the conduct of the scientific work under the United States Government, and to report the result of its investigations to Congress. In order that the subject may be clearly understood, the language of Section 8 of the act referred to should be quoted:

" SEC. 8. The National Academy of Sciences is required, at their next meeting, to take into consideration the methods and expenses of conducting all surveys of a scientific character, and all chemical, testing, and experimental laboratories, and to report to Congress as soon thereafter as may be practicable a plan for consolidating such surveys, chemical, testing, and experimental laboratories, so as to effectually prevent duplication of work and reduce expenditures without detriment to the public service."

A committee was promptly appointed, and that committee gave serious and prolonged attention to the subject. In due time the committee submitted its report to the Council of the Academy. The Council having approved, the President transmitted the report to the Speaker of the House of Representatives and the presiding officer of the Senate. Everything was done in proper form, so far as could be determined. The President of the Academy congratulated himself on the personnel of the committee which he had appointed, upon the report, and upon the fact that the Academy had performed an important duty and had been, as he thought, of real service to the National Government.

It were well, perhaps, to close the account of the incident at this point, but unfortunately the moral would be lost, and the only object of telling the story at all is to point the moral. Well, what happened next? It is not necessary to go into detail. The result was humiliating to the committee that drew up the report—and possibly to the President. That report seems to have been promptly pigeonholed. It is certain that, so far as we have any information on the subject, it was not given serious consideration by Congress. And yet, whatever may have been its imperfections, that report represented the views of a group of eminent men of science who had devoted much time and thought to the study of the problem before them and who, at the request of the President of the United States, had been given every opportunity to learn the facts. Such an experience need not dishearten. The charter still holds good; and, accordingly, the

Academy stands ready, whenever called upon by any department of the Government, to " . . . . investigate, examine, experiment, and report upon any subject of science or art. . . . ."

As time passes, it will come to be recognized more and more clearly by those in authority that the scientific method is the one most likely to lead to results of permanent value. Briefly defined, the scientific method consists in studying the facts and then drawing the most logical conclusion from these facts. It is most desirable that our Government should utilize to a greater and greater extent this method which is free from partisanship and has only truth to serve. In the long run, the influence of the National Academy upon affairs of government must be felt.

Farsighted statesmen must see, and do see, that it is well for the country to have a body of workers in the field of science connected in some way with the Government, and the day will come when this will be recognized more clearly and more generally than it is today. The question is not what is best for the Academy; it is, what is best for the country. May we not hope that in the near future Congress will see its way clear to emphasize the importance of the connection between the Government and the Academy by providing it with a proper home which can serve as a center of general scientific activity? This subject has again recently been brought to the front, and there is a possibility that favorable action may be taken.

By an act of Congress approved June 20, 1884, the National Academy of Sciences was " . . . . authorized and empowered to receive bequests and donations and hold the same in trust, to be applied by the said academy in aid of scientific investigations and according to the will of the donors."

The funds under the general management of the Academy, and their purposes, should receive some notice at this point, in order that the work of the Academy may be more clearly understood. I will go over this briefly. Some of them will be referred to at the time of the award of medals tomorrow afternoon. I do not wish to say anything that will necessitate repetition at that time.

First there is the A. D. Bache Fund. This amounts to over $50,000. It was provided by the will of Alexander Dallas Bache, one of the charter members, and the first President of the Academy, who was for many years superintendent of the United States Coast Survey. The Academy is trustee, and the income is applied to the prosecution of researches in physical and natural sciences.

Second, the Joseph Henry Fund. This fund of $40,000 was contributed by a number of friends and admirers, as an expression of the donors' respect and esteem for Professor Joseph Henry's personal virtues, their sense of his life's great devotion to science with its results of important discoveries, and of his constant labors to increase and diffuse knowledge and promote the welfare of mankind. The income was to be paid to Professor Henry during his life, and after his death to his wife and daughters, and after the death of the last survivor, the fund is to be delivered to the National Academy of Sciences; the principal to be forever held intact, and the income to be from time to time applied by the said National Academy of Sciences in its sole discretion to assist meritorious investigators especially in the direction of original research. Happily, this fund has not yet come into the possession of the Academy. It is not necessary to remind this audience that Professor Henry was for years the Secretary of the Smithsonian Institution.

Third, the J. C. Watson Fund. This amounts to $25,000, and was provided by the will of Professor J. C. Watson, a distinguished member of the Academy, who died in 1880. The income shall be expended by said Academy for the promotion of astronomical science. It is also provided " . . . . that the Academy may if it shall seem proper provide for a gold medal of the value of one hundred dollars to be awarded . . . . from time to time, to the person in any country who shall make any astronomical discovery or produce any astronomical work worthy of special reward as contributing to our science." Five medals have thus far been awarded, the recipients being B. A. Gould, Edward Schönfeld, Arthur Auwers, Seth C. Chandler, and Sir David Gill.

Fourth, the Henry Draper Fund. This is the fund which makes possible the award of the Henry Draper Medal, to be awarded tomorrow afternoon, with, no doubt, some explanatory remarks. The Draper medallists named in chronological order, are S. P. Langley, E. C. Pickering, H. A. Rowland, H. K. Vogel, J. E. Keeler, Sir William Huggins, G. E. Hale, W. W. Campbell, and C. G. Abbot.

Fifth, the J. Lawrence Smith Fund. In 1884 Mrs. J. Lawrence Smith, widow of one of our honored members, presented to the Academy the sum of $8000, the object of the gift being to promote the study of meteoric bodies, a branch of science which Dr. Smith had pursued with marked success. In accordance with the wishes of the donor, it was decided that a gold medal to be given as a reward for original investigations, would be most appropriate. Any excess of income above what is necessary for the striking of the medal, " . . . . shall be used in such manner as shall be selected by the National Academy of Sciences in aid of investigations of meteoric bodies to be made and carried on by a citizen or citizens of the United States of America." Only one J. Lawrence Smith Medal has been awarded. The recipient was H. A. Newton, and the medal was awarded for the investigation of the orbits of meteors. The income has otherwise been used to aid investigations, especially those of Professor Newton.

Sixth, the F. A. P. Barnard Medal. This medal is not provided for by a fund which was given to the Academy, but the Academy has the duty of naming the medallists. It is given every five years. It is not to be awarded this year. It was awarded in 1895 to Lord Rayleigh and William Ramsay. By the way, the money was left to Columbia University, of which Professor Barnard was the President at the time of his death. The medal is awarded for meritorious services to science, to such person, whether a citizen of the United States or of any other country, as shall, within the five years next preceding, have made such discovery in physical or astronomical science, or such novel application of science to purposes beneficial to the human race, as, in the judgment of the National Academy of Sciences of the United States, shall be esteemed most worthy of such honor. In accordance with these terms, the Academy has recommended to the trustees of Columbia University awards of the Barnard Medal as follows:

In 1895 to Lord Rayleigh and William Ramsay—now Sir William Ramsay—for their brilliant discovery of argon, a discovery which illustrates so completely the value of exact scientific methods in the investigation of the physical properties of matter.

In 1900 to Wilhelm Conrad Röntgen, for his discovery of the X-rays.

In 1905 to Henri Becquerel, for his discoveries in the field of radio-activity.

In 1910 to Ernest Rutherford, for meritorious services to science resulting especially from his investigations of the phenomena of radio-active materials.

Seventh, the Wolcott Gibbs Fund. When Wolcott Gibbs, who was one of the incorporators of the Academy and at one time its honored President, reached the age of seventy in 1892, a number of friends presented him with a sum of money to establish a fund bearing his name, the income to be devoted to aiding in the prosecution of chemical research. Dr. Gibbs presented this fund to the Academy, the income to be administered by a board of directors, who ". . . . shall have absolute and entire control of the disposition of the income of the fund, employing it in such manner as they may deem for the best interest of chemical science."

Eighth, the Benjamin Apthorp Gould Fund. This was given by Miss Alice Bache Gould, daughter of the distinguished astronomer, Benjamin Apthorp Gould, one of the incorporators of the Academy, who died in 1896. The amount was $20,000, given as a memorial of the life work of her father, the income to be used for the prosecution of researches in astronomy.

Ninth, the Cyrus B. Comstock Fund, which came into our possession in 1907. The prize in money has not been awarded thus far, but will be awarded at the meeting at the White House tomorrow, and explanatory remarks made at that time. General Comstock was a distinguished engineer and member of the Academy, who died in 1910.

Tenth, the O. C. Marsh Fund. Professor Marsh, for twelve years President of the Academy, died in 1899. He bequeathed the sum of $10,000 to the Academy, the income to be used and expended by it for promoting original research in the natural sciences. This fund has not yet become available.

Eleventh, the Alexander Agassiz Fund. Alexander Agassiz, who was President of the Academy from 1901 to 1907, died in 1910, and bequeathed to the Academy the sum of $50,000, unconditionally. No decision has yet been reached in regard to the uses to which this fund is to be put.

And, finally, the Agassiz Medal, which will be awarded for the first time this year, was provided for by a gift of Sir John Murray.

While this account may have proved tedious to some of you, it seemed necessary for the purpose of giving a correct impression of the work being carried on. The Academy has sacred duties to perform. It will soon devolve upon the younger members to see that these duties are conscientiously performed.

The constitution provides that the Academy shall hold one meeting each year in the city of Washington and another at such place and time as the Council may determine. Whatever may be said of the duties of the Academy as the scientific adviser of the Government, and as a custodian of trust funds, it must be acknowledged that it is through the agency of its regular meetings that its influence is mainly exerted. In this, as in other matters, it is the subtle, the intangible, the spiritual that tells.

Workers in the field of science are supposed by some—perhaps by many—to be incapable of recognizing the force of the intangible, and yet scientific work must inevitably lead to this recognition. It is impossible to weigh and measure the effect of the meetings upon those who take part; but that effect is felt none the less, and it is certain that those who attend are in the long run benefited, some in one way, some in another.

This is not a subject that lends itself to profitable discussion. It may not be out of place, however, for one who has been a regular attendant for more than thirty years to make public acknowledgment of the debt which he personally owes the Academy for the opportunities it has afforded him of associating with and counting among his friends those whose earnest, honest work has been an inspiration to him and to the world. This association has been an inestimable privilege, for which he is deeply thankful.

The work of the Academy will continue; new and younger members will take up the work. Is it too much to hope that when the centennial anniversary is celebrated some of the members here present may be remembered as we today remember, with gratitude, our founders? (Applause.)

THE PRESIDENT: The first speaker this morning, as announced on the printed program, does not need any introduction. In fact, none of the speakers who will appear here need introduction.

The first speaker is President Hadley, of Yale. I may add to what has already appeared from what I have said, that he is the son of one of the early members. He will speak on " The Relation of Science to the Higher Education in America." President Hadley. (Applause.)

ADDRESS OF PRESIDENT ARTHUR T. HADLEY
OF YALE

ON

"THE RELATION OF SCIENCE TO THE HIGHER EDUCATION IN AMERICA"

DR. HADLEY: *Mr. President, Members of the National Academy, and Ladies and Gentlemen:* The half century which has elapsed since the founding of this Academy has witnessed, as we all know, a radical change in the relations between science and education. This change is equally marked in professional training which prepares students for their several callings, and in the general training which prepares them for the duties and enjoyments of citizenship.

Fifty years ago the professional study of science in our universities was confined within very narrow limits, surprisingly narrow to those who see those places as they are today. There was no room for science in the schools of theology or of law. Schools of philosophy, in the modern sense of the word, had hardly developed. Even in schools of medicine, where among universities the study of natural sciences first gained a foothold, there was relatively little of scientific method as we today understand the words, either in the teaching or in the study. There was much more learning of names of things and much less learning of reasons of things; much more of tradition and much less of investigation. The anatomy and chemistry of the medical schools of those days were good sciences, as far as they went, but they generally did not go very far. As to the use made of it, there is truth in the remark of one of my former colleagues that down to a recent day the three learned professions of theology, law and medicine had not advanced far beyond the old conception of the magic of the tribal medicine man, that the important thing for science to do was to find proper formulas of exorcism with which to banish evil spirits from their several realms of action.

Outside of the universities, a half century ago, things were little or no better. There was a small number of schools of engineering and a still smaller number of schools of chemical technology, but they did not form part of a large scheme of business training as a whole for the nation. The trained engineers of the first half of the nineteenth century were, as a rule, engineers trained in military schools. Most of the civil engineers, as we called them by contrast in those days, had learned their profession in the field. Most of the technologists had learned it in the shop. Now, all of this has changed during the fifty years of the life of the Academy, and changed radically. Our universities have developed scientific study in all their departments, and especially so in their schools of medicine and philosophy. And side by side with these universities, schools or faculties there have grown up colleges of engineering and technology, sometimes in connection with the university, sometimes outside of it, which lay a scientific foundation for a calling that only a few years ago was thought to need no scientific foundation at all. For the world has found a place for the scientific expert in every line and is inclined to regard as the best school, not the one that has the most students, not even the one that can give the best general education, but that which in the different lines can train and furnish scientific experts of the highest rank and most varied knowledge.

For civilized nations have at last come to the conclusion that the old supposed antagonism between theory and practice was a misleading conception, and the habit of drawing a sharp line between the theoretical man and the practical man was a pernicious one.

Fifty years ago a man who had obtained all his knowledge of his business by his own experience was habitually proud of the fact; he was, as the phrase went in those days, a self-made man who spent most of his time in worshipping his creator. (Laughter.) He counted it a matter of superiority that he knew nothing except what he had found out himself and taught himself. Today it is recognized that every man can learn from the theorist that there is room for the application of scientific principles in every department of life; that the farmer, the manufacturer or the merchant, no less than the engineer or the physician, must prepare to avail himself of the theory which has been built up by investigators, which has been taught in laboratories and incorporated in books, if he would bring his practice up to the needs of the time.

Of all the conquests of modern science, there is none which, in my judgment, is more remarkable or significant than this conquest of current business opinion. We no longer draw a distinction between learned and unlearned professions. We have recognized that every profession and every trade, in order to be pursued to the best advantage, must be a learned one. None so complex as to be unable to get help from science; none so simple as not to need it. We have shaped our system of technical training accordingly; and we have learned to rate at their true worth the men and the places that can give training as research institutions, side by side with universities which make progress in such training possible.

Equally important, though of a different and perhaps less wholly satisfactory character, has been the change in the scheme of our general education, in the choice of subjects and methods of teaching offered in preparation for the general work of citizenship as distinguished from the preparation of each man for his business or calling.

The old course of study in our high schools and colleges consisted chiefly of classics, mathematics, and metaphysics, with a little history and a few descriptive courses in natural science. Of scientific training in the modern sense it gave none, except to the unusual man whose mathematical tastes made the study of algebra and analytical geometry a means of scientific education in spite of text book and instructor or class room atmosphere, or the still more unusual man who used his grammar and metaphysics as an exercise in closely ordered reasoning. The course as a whole was constructed for the man whose interests were in the past rather than in the present and the future. The training which it gave—good, in many respects—was a training in memory, in expression and in accuracy of apprehending language, one's own or another's, rather than in scientific method as we understand it today.

There is on the façade of the main hall of a university which has done much for education in many lines, a representation of Philosophy in a dominant central position—old fashioned metaphysical philosophy—with the different sciences laying tribute at her feet. I suspect that this is not an unfair characterization of the views as to the place of science in education which prevailed among most college faculties a generation or two ago.

Now let me say right here that I do not for a moment overlook the advantages of the old system. It taught the boys to use books and find things out from books, and to expect to do hard work for that purpose instead of to have somebody else make it easy. This was a great merit, and the boys trained under the old system showed this merit. But college faculties were often blind to the particular kind of book learning and to the general kind of book learning that was most important for human progress and which was of most concern to the living world outside.

For at the time when the Academy was founded, and in the time since, chemistry and physics and geology and biology were becoming not only matters of importance to the experts in their several callings, as I have indicated, but matters of real and dominant interest to intelligent men who were not experts, but who cared for knowledge and who cared for current history. A large section of the world, an increasingly large section of the world, cared more for books that explained the tendencies of the present than for those that embodied the ideals of the past. Perhaps this movement may have gone too far and may have caused people to care too little for the ideals of the past, to overvalue scientific reading as compared with historical or literary reading. I shall not try to discuss whether it did or not. At any rate, a curriculum which was exclusively occupied with classics and philosophy did not fully meet the demands of grown men or the needs of boys, and inevitably the course of study in our colleges had to be remodelled accordingly.

Each decade of the last fifty years has witnessed a gradual crowding out of classics from our older schools and colleges by subjects of new and more present interest, and a growth of new schools and colleges of a different kind, where science in varying forms is made the chief subject of attention and other matters relegated more or less to the background.

Now this increasing interest in science is a matter about which we may all, members of the Academy, and guests of the Academy, scientific men and literary men, rejoice heartily. But how far the things that are called science deserve the name of science, or how far the teaching of such subjects by present methods always deserves the name of education, is quite another question. Every school superintendent likes to stimulate the attention of his pupils by giving them the opportunity to see amusing phenomena with their own eyes, and if possible set them in motion with their own hands. Under some circumstances this may be the best kind of scientific training; under other circumstances it may be no training at all.

Nature study—to quote a phrase which is popular among educators of the present time—is good if it is made the basis for teaching scientific methods, and bad if it is simply made a means of momentary amusement. Unfortunately, a large part of our school committees and school teachers think that the subject makes the science. They may not go as far as the author of a reputable work, " Murray's Handbook to Spain," who says that the mountains of that country, to quote his own words, " abound in botany and zoology." They are apt to assume that the picking to pieces of flowers is in itself botany, and that hearing a carbon disulphide mixture make a loud explosion, marks progress in chemistry, and to act accordingly.

Fifty years ago the members of the National Academy of Sciences who held seats in college faculties were occupied in protecting science against its enemies. I am not sure but what today their chief duty lies in protecting it against its friends. (Laughter.)

3

When the National Educational Association says that high schools should be encouraged to omit the study of algebra and geometry and that the colleges should be compelled to accept for admission an equivalent amount of " science "—God save the mark—it is time for the true friends of science to call a halt.

I may say in parentheses that only the other day the school superintendent in one of our more newly developed parts of the country said he had to make a change because it was so easy to find thoroughly competent teachers of physics, but that so few of them ever knew any algebra. (Laughter.) For the importance of scientific training to the student in our high schools and colleges is not due primarily, or in large measure, to the facts of physics or biology that he learns in the school. It is due to the training in certain habits of observation and deduction in certain methods of hypothesis and verification, which he can get more effectively by a good course in science than by one predominantly devoted to languages, where the scientific training is merely incidental. That the facts of physics or biology are more interesting to the student of the world than those of Latin and Greek and have more obvious bearing on everyday life is a help to the teacher in securing the voluntary co-operation of the pupil; but it is far from being the fundamental reason why the subjects themselves are educationally valuable. It is not the subject that makes the course scientific; it is the method.

You have been good enough, Mr. President, to refer to my father's connection with the Academy, and I for my part am glad to take the opportunity to say that he regarded his election to membership in this body as the greatest honor he ever received. I feel sure, therefore, that I shall be pardoned if I illustrate the point I have just made by reference to my father's teaching.

Fifty years ago the one course in the Academic Department of Yale College, where modern science was really taught, was the course in Freshman Greek. For my father, though he had the highest enjoyment of classical literature, was, by training and temperament, a philologist; and he taught the Freshmen who came under him to take Greek verbs to pieces and compare and observe their parts and put them together again, and see what principles were involved in the analysis and synthesis, exactly as the botanist may have done with his plants or the chemist with his elements.

In those days chemistry and physics were taught in Yale College, as distinct from the Sheffield Scientific School, solely by text books and lectures. Philology was taught by the laboratory method, and for that reason the Freshman Greek course was a course in modern science and meant that to the pupils. (Laughter.)

The courses in chemistry and physics widened the boys' knowledge of facts and doubtless encouraged many of them to get scientific training for themselves afterward; but the course in Freshman Greek, for the first term, was a course in science, because the boys learned to do the things, both easy and hard, which are the heritage of the man of science.

Science is not a department of life which may be partitioned off from other parts; it is not the knowledge of certain kinds of facts and the observation of certain kinds of interest, as distinct from other facts and other interests; it is a way of looking at life and dealing with life; a way of finding out facts of every kind and dealing with interests as varied as the world itself,

" Where each for the joy of the working, and each in his separate star,
Shall draw the thing as he sees it, for the God of things as they are."

(Applause.)

THE PRESIDENT: I now present to you the next speaker, who will discuss " International Cooperation in Research." The speaker is Dr. Arthur Schuster, well known to many of us, whose title is here given, though he has many, as the Secretary of the Royal Society of London, a society which I may remind you celebrated the two hundred and fiftieth anniversary of its foundation last summer—the mother of all scientific associations. Dr. Schuster. (Applause.)

ADDRESS OF DR. ARTHUR SCHUSTER
Secretary of the Royal Society of London

ON

" INTERNATIONAL COOPERATION IN RESEARCH "

DR. SCHUSTER: *Mr. President, and Ladies and Gentlemen:*
The intellectual activity of the world, scientific, literary or
emotional, passes alternately through fertile and through barren
periods. Each fertile period has its characteristic peculiarities
and though any one generation may not be competent to form a
just estimate of its powers and effects, it is able to compare the
fruits of its own labors with the harvest of its predecessors. You
will probably agree with me that our age is distinguished by
having disclosed a vast array of facts which take us nearer to the
infinitesimal structure of matter and which reach further into
the infinite design of the universe than the boldest flight of
imagination could have foreseen half a century ago. But we do
not flatter ourselves that the intellect of our time, judged by the
power of individuals, is exceptionally great. No doubt, men
of commanding genius are still with us, but they are not more
numerous or more original than in former times. What then
is the peculiarity that has produced such great results? In my
opinion what has been accomplished is due in great part to the
spread of higher education, which has evolved an army of
competent investigators possessing enthusiasm for research
which now, for the first time, is led into useful paths by the
few great minds, whose powers thus receive a wider range and
become more productive. It is in this that our great strength
lies.

The functions of an organization devoted to research are to
take full advantage of all available mental resources. Intellect
can not be artificially created nor can originality be taught, but
whatever intellect and originality exists may be directed into

fertile channels, so that those who have the gift of connecting facts shall not fail because the facts are not available.

The advance of science demands that experiment or observation and theoretical discussion should advance in parallel lines. Without organization, one of the teams on whose joint exertions the advance depends, is likely to outrun the other. Thus Newton, when he had formulated his law of gravitation, which connects the orbit of the moon with the acceleration of falling bodies, did not publish his discovery for many years, because he could not verify his theory as closely as he desired. It was only after the French Academy had accurately measured an arc of meridian and had discovered a substantial error in previous measurements that Newton's law of gravitation could be said to be proved. In this case theory had gone ahead of observation; but examples of the opposite kind will not be wanting so long as we have observers concerned entirely with the accumulation of data, content to leave discussion to the dim future. It is one of the objects of organizing science to bring the two factors to bear on each other.

International co-operation in research is necessary because scientific inquiries can not be divided into compartments limited by political boundaries. The very language which we use to express our thoughts is tied down by conventions, some of which we have absorbed as students, but which in the case of new branches of learning have formally to be agreed upon. Our measurements—and all accurate science depends on measurements—have to be expressed in units, and how are these units to be fixed except by agreement? While this will be acknowledged by everyone, it is not equally recognized how much our present refinements in scientific research depend on organized efforts. Whether these efforts should be concentrated in a single laboratory or confined within one political unit or carried out by the combined scientific community of the world, mainly depends on the nature of the problem.

It is not my purpose to trace in detail the history of international problems and international organizations; but, rather, to show the great variety of problems in which useful results have already been achieved by international co-operation, and to bring the lessons of the past years to bear on the future.

I divide international co-operation into three categories:

1. Agreement on standards and units of measurement.

2. The distribution of work bearing on the same problem, between different nation's, for the purpose of economizing time and expenditure.

3. The investigation of problems which can not be solved unless observations made with identical or similar instruments are obtained from different parts of the world and the records published in a homogeneous form.

I think all are agreed as to the question of units and I need not detain you by giving you an account of the various international conferences which have been held and agreements which have been arrived at on these matters.

As regards problems of the second category, they are those which deal mainly with the cosmos as a whole because their solution depends so much on the collection of statistics which exceeds the powers of individuals or even of single nations. A few examples may illustrate what has already been accomplished. First, and foremost, we have the great Star Catalogue, initiated at an international congress twenty-five years ago, when eighteen observatories combined to divide the work, each taking a number of zones in the heavens.

The importance of this work will be plain to everyone, and we must regret that it is still so far from being completed.

As it is not my intention simply to point out the merits of international work, but also to point out its difficulties, a few words may be said which are not intended as criticism, but which may serve to point out the weakness which arises when there is no central authority which lives longer than the single individual can expect to live.

Pioneers will always be found to initiate a work, but in time they die or retire from office; others take their places, and if these become more interested in fresh problems, the work suffers unless it is effectively impressed on their attention by some permanent body. Where to find such a central body, whose main functions would be to endow an undertaking with sufficient inertia to carry it over periods in which the work may seem to be a drudgery, is a matter which deserves careful consideration.

The completion of the Star Catalogue, which has given rise to these remarks, is only the beginning of an even greater piece of work. When we have determined the positions and magnitude of stars at any one time, we have only taken the first step towards solving the main problem, and must proceed to measure the proper motions, the parallaxes, and also map the spectra. This work is so vast that all hope to accomplish it within reasonable limits is difficult and has to be abandoned unless our statistical ambitions are lowered, and instead of taking the complete sphere of the heavens we select restricted but typical areas for detailed examination. This has been done on the initiative of Professor Kapteyn, who has secured a sufficient number of voluntary associates who are now carrying out a combined undertaking which has already yielded results of the greatest importance, and you will hear something more of this work from his own lips.

Now the essence of work of this kind consists in shortening the time required to accomplish an extensive task by dividing it among a number of persons. If the work is purely statistical, it may be complete in itself, and the published records become then available to anyone who requires them. In other cases, the observations may have to be collected by a central authority and treated by recognized methods of statistics or analysis before they become useful to the scientific public. While it is generally the observational portion of the work that is subdivided and the discussion that is centralized, the reverse is the case in the proposal made by Professor Pickering—that one central observatory in a favorable position should furnish photographs in sufficient numbers and distribute them among astronomers all over the world, to be measured and discussed.

Finally, a great undertaking of quite a different character—the International Catalogue of Scientific Literature—must be classed in the same category. This catalogue has arisen out of a desire to classify the scientific literature of the world, so as to enable anyone who desires to study a certain subject, to find out quickly all previous researches relating to it. Practically all nations in which scientific work is carried out have united, each collecting its own data and forwarding it to the central bureau in London.

I can not pass away from this type of international co-operation without expressing regret that a proposal which was made by the late Professor Simon Newcomb has not been adopted hitherto. When the first program of the Carnegie Institution of Washington was being discussed, he proposed that there should be some central computing bureau established at one place where accumulated data of observation, which required scientific treatment, could be discussed and treated in that way. The number of instances which have come to my own notice within the last few years, in which the existence of such a bureau would have been of the greatest assistance to the progress of science is considerable; and I feel very little doubt that others have also felt the want.

The problems which fall into the third category are mainly those belonging to the important and much neglected subject of geophysics. The time is passed when we could separate the physics of the laboratory from that of the earth, and that again from the physics of the universe. The experimenter who now studies the structure of the atom must keep an eye on the sun and stars in order to detect whether celestial observations destroy his theories or give them strength.

Atmospheric electricity and terrestrial magnetism, treated too long as isolated phenomena may give us hints on hitherto unknown properties of matter. A meteorologist, finding out at last that space has three dimensions, and that the motion of air is governed by the laws of mechanics, has converted what hitherto has been a sport into a science.

Before enumerating the international associations which are dealing with these problems of geophysics, let us say a few words as to the problems themselves.

We have, first, to study the shape of the earth and the variations in the gravitational forces which are observed on its surface. We have further to take account of the secular variations of level and of the more or less violent disturbances which accompany earthquakes and earth tremors. By comparing the indications of instruments placed in different localities, we can deduce the rate of propagation over the earth and through the earth of the seismic waves. This yields us important information on the physical properties or material composing the interior of the earth. The cause of terrestrial magnetism is at present unknown, and we have no means at our disposal to attack the problem directly, but the study of the diurnal and secular variations, may give us a clue, and deserves our closest attention.

In a similar way, the study of the higher atmosphere and of the high electric conductivity which the air is now known to possess at heights which we can not reach, is also a subject which can only be studied by combined efforts. How are these questions dealt with at present?

We have, first, an International Association of Geodetics, which is an exceedingly efficient body, with a bureau at Potsdam, under Professor Helmert. That association is successful, perhaps, partly because its work has been facilitated in that it had to build on virgin soil. Nothing had been done, to a very great extent at any rate, internationally before that association came into being. On the other hand we have the International Association of Seismology, a related subject, which was only founded at the beginning of the present century, with a central bureau at Strasburg. This association had to overcome more serious difficulties. It entered into the field when there was already a less extensive organization in existence, which had been originated by Professor Milne and was directed by a committee of the British Association. The question of instruments also presented peculiar difficulties, which it is hoped may soon be overcome.

As regards terrestrial magnetism, I have only a few words to say.

Through the magnificent efforts of the Carnegie Institution of Washington, we are at last likely to have a satisfactory magnetic survey of the world, but important as the results obtained by Professor Bauer in the " Carnegie " will prove to be, they will have to be supplemented by systematic observations of the variations of the magnetic forces at a number of fixed stations. Many such stations are in existence, though they are very irregularly distributed over the surface of the earth.

In this subject, almost more than in any other, an international agreement on the manner in which the records are to be treated and published is essential, and it is much to be regretted that the attempts that have been made to reach such agreements have not met with greater success. There are, no doubt, peculiar difficulties due to differences in the organization of the magnetic services.

Methods have developed independently in different countries, and there is a natural but regrettable reluctance to alter an instrumental detail or a peculiarity in treating the observation until the necessity of the change has been demonstrated. But that can never be done, because practically all methods are equally good. What is bad is that they differ. Almost any one of these methods could be adopted with advantage anywhere; so that a discussion of which of the methods is better than the other is futile. The first essential then is that in every place on earth the same methods should be adopted, because the least difference in them may cause important errors in the deductions when they come to be compared with each other.

The only body which at present deals systematically with the records of terrestrial magnetism is a sub-committee of the Meeting of Directors of Meteorological Observations. The Directors of Meteorological Observations meeting at intervals have appointed a certain number of sub-committees dealing with a certain number of subjects. Some of these overlap other associations already. So that, for example, the question of solar radiation falls partly under that sub-committee of the Directors of Meteorological Observations and also under the International Solar Union, a union which has been founded by your Foreign Secretary, Professor Hale.

The present international organizations differ considerably in the manner in which their expenditure is provided for. The International Geodetic Association—the Association of Seismology—and the International Bureau of Standards, are directly supported by the governments, the contributions depending upon the population of each country and amounting, for the larger ones to—I need not give you the figures now. They are of no particular interest.

The International Catalogue of Scientific Literature is a very costly undertaking, and that is provided for by each country guaranteeing the sale of a certain number of copies; a capital fund having been paid to start the organization by the Royal Society of London.

In the case of the Great Star Catalogue, each observatory is responsible for its own expenditure. The four French observatories have received government contributions amounting together to over $500,000. In England a much smaller sum has been given, and in other countries the work has languished a good deal because sufficient funds were not available.

The Solar Union has no funds whatsoever and is even unable to pay for its own publications. Sufficient has been said to show how wide a range is already covered by international research. Further extensions of the work are constantly being called for, and we are brought face to face with the problem that separate associations can not be multiplied indefinitely without introducing difficulties which, as their number increases, endanger the objects which they are intended to serve. Apart from the overlapping of interests and questions of finance, the time spent in correspondence and administration is already serious. The nature of the problems suitable to be dealt with by international efforts is such that the same persons are generally interested in several of them, and the meetings succeed each other so rapidly as to become a serious tax on the time of those who attend them and some who used to look with favor on international work are beginning to be frightened.

Perhaps we may look forward to some arrangement to combine the meetings of the different associations in the manner of the different sections of the British Association, for instance. But this would require some central authority to act as a bond between the bodies which at present are separate and independent.

Economy of working, both financial and administrative, points in the same direction, and we are driven to the conclusion—and that, I think, I should like to make the moral of this paper— that the present policy of establishing a separate association for each new extension of international work should be reconsidered and an effort made to economize time in working and administration by some larger scheme including the various separate international institutions on related and similar subjects.

Realizing that it is necessary to take some action in this direction, yet perhaps not understanding correctly why the action is necessary, an ambitious undertaking has been evolved in Belgium, where it is proposed to erect an office uniting all international associations, whatever their object or character may be. The promoters have drawn up their statutes, one general congress has already been held, and another is now being organized. No success can, however, be expected from a scheme launched by a self-constituted and irresponsible body, unless its program commands general respect.

Is this the case in the present instance?

I do not know whether you realize the number of associations which exist. I shall not call them " international associations," but associations which call themselves international. (Laughter.) The number to be united in this Belgium scheme is 279, and each of them, if I understand the proposals correctly, may have an office in a large building to be erected for the purpose. If you read through the list of these associations, I do not know what your feeling would be, but I can describe to you what mine has been; and it is exactly like that which I should have if I were to enter a museum, and find, side by side the Venus of Milo, a living tiger, a collection of rare manuscripts, and sanitary appliances. (Laughter.) You will be interested to hear that, amongst the institutions which are to be provided for in this building, is the International Bureau of American Republics; but it is also intended to include " The International Congress for Providing Cheap Lodgings." Anyone who enters the building and tries to find the particular room to which he wants to go has to ask the man in charge. You can imagine this kind of a conversation taking place:

" Is this the International Union of Friends of Young Girls? "
" No, but it is the International Congress of Commercial Travelers."

(Laughter.)

The architect, no doubt will do his best to group together associations relating to the same subjects, and it would be interesting to pass through the corridors devoted to all the religious and irreligious societies that take the name of " international." If a humorist were to exchange the name plates over the doors, the mathematician who has traveled all the way from Australia to attend the " International Congress for Promoting the Study of Quaternions " might find himself in the room reserved for the " International .Union of Woman Suffrage," and a member of the " Association of Seismology " might be mixed up with the association to prevent the abuse of alcohol. (Laughter.)

I do not like to throw ridicule on what is obviously a well meant effort, but however much our sympathy may extend to each of these objects separately, no good purpose is served by inventing a connecting link between incommensurate objects, such as solar research and the proper observation of Sundays. (Laughter.)

Our work is sufficiently difficult, if we confine ourselves to scientific methods. It nevertheless remains true that it is desirable to establish some central authority which can act as a connecting link between different associations. What should its functions be? It is the essence of all international combinations that they depend entirely on moral force and have no power to impose their decisions. A central authority must therefore be content with offering advice, with the conviction that, if the advice is sound, it will be accepted.

Though the existing associations would tolerate no interference with their independence, they would doubtless consider with care any suggestions made to them in the interests of science by an authoritative body. Our problem is therefore to find an authority of sufficient eminence to be generally looked upon with confidence and who could also act as adviser to different governments when they are asked to financially support some fresh undertaking. That is one of the most serious difficulties of the present time. There is a new international undertaking proposed almost every year, and application is made to the different governments for support and money. What is the government to do? To whom is the government to go for advice whether such an undertaking is worthy of support or not? My solution of that question is this: In the International Association of Academies we possess indeed a body fulfilling all the requirements of such a central authority, provided the individual academies constituting the association are willing to undertake the task. The Association of Academies was founded at a conference held at Wiesbaden on the 9th and 10th of October, 1899, the National Academy of the United States being represented by Professors Newcomb and Bowditch. The paragraph of its statutes which were adopted at a meeting held in Paris in 1901 relating to the functions of the association runs as follows:

"The object of the association is to prepare and promote scientific work of general interest which has been submitted to it by one of the associated academies, and to facilitate in a general manner scientific intercourse between different nations."

From its origin the association claimed an advisory voice in new international undertakings, and at the meeting held in London in 1904, the following resolution was passed with one dissenting voice:

" That the initiation of any new international organization to be maintained by subventions from different states demands careful previous examination into the value and objects of such organizations, and that it is desirable that proposals to establish such organization should be considered by the International Association of Academies before definite action is taken."

After a period of activity ranging over about twelve years it may be useful to review the work which has been accomplished, but I shall confine myself to the record of its section of science, remarking only that the section of letters has also much important work in hand.

The powers of the association are purely advisory; it has no funds at its disposal and for this reason alone is unable to initiate or support any scientific enterprise unless the individual academies provide the expenditure, as is being done, for instance, in the publication of Leibnitz's works, which has been undertaken by the Academies of Berlin and Paris jointly. A complete map of the moon with its features named according to an agreed scheme is in process of preparation and is welcomed by students of the lunar surface. Among the subjects which have been treated, the excellent work done by an autonomous committee appointed to investigate the functions of the brain should also be referred to; and there are a number of various committees which have done good work.

In many cases the association has been called upon to express a favorable opinion on the importance of some international scheme which is independently being pressed upon the consideration of one or more governments. To deliver a platonic blessing is so gratifying a task that applications for it are not perhaps always scrutinized with sufficient care, though I admit that it is better to support a doubtful enterprise than to risk stopping a good one.

The association has been most successful when it has used its influence to press important scientific objects on the attention of their governments. It is in part at any rate due to their recommendation that money was found for the measurement of the great arc of meridian, which, covering 105 degrees, stretches through Russia and Roumania and continues through Asia Minor and Western Africa, to the Cape of Good Hope. This is a continuous arc of meridian reaching from the north of Russia to the Cape of Good Hope in which a number of governments, the British Government, the German Government, the Russian Government and the Turkish Government are involved, which is in process already, and is really nearing completion.

It has become the practice during recent years that international organizations established independently place themselves under the protection of the Association of Academies, to which they report periodically. Though the academies exercise no control over such bodies they stand to them as a reserve power willing to help when required.

In all these respects the association has fulfilled the intention of its founders, but has it left its mark to any appreciable extent on the progress of science? Without wishing to underrate the good that this body has done in the past I do not think I stand alone in hoping for a wider activity in the future, and I doubt whether it will long maintain its vitality unless it extends its ambitions as it passes from the age of youth to that of manhood. This is a critical period in its history, and much will depend on the policy it will adopt on a question which may still be kept in abeyance for a short time, but which will have to be faced before long.

An international organization which has no central office and is not domiciled in any country is not a legally constituted body. It possesses no property. It cannot accept gifts or legacies. The question has been repeatedly raised whether it is desirable to remove this restriction and to establish the association on a legal foundation. For this purpose it would have to place itself under the laws of some one country, and the selection of that country complicates the decision on the main issue, as national consideration and perhaps to some extent national jealousies have to be taken into account.

To clear our minds, let us separate the two issues, that of the power to hold property and that of a permanent domicile. Each academy knows from its own experience that though individual research may often be carried out at a small cost an organized investigation demands funds which become considerable when its range is wide. It is therefore just the type of work that an international organization is best fitted to undertake which demands the greatest amount of assistance.

The question to be faced is this:

Shall our International Association be forever content to exercise a purely platonic patronage, or shall it take an active part in promoting research? If it chooses the latter course it seems to me to be indispensable that it should have funds at its disposal.

I advocate the bolder policy on two grounds: Firstly, international research is most logically administered and paid for by international funds and, secondly, it seems to me that a purely moral support can not in the long run remain effective. The existing special associations, as I have already stated, must retain their complete independence, and it is not likely that it will ever be desirable that the Association of Academies should undertake any work in which financial support is expected to extend over a considerable period; but when promising enterprises are in their experimental stages, funds are often most urgently required and most difficult to obtain.

It is here that an international body, having an independent income, could most efficiently step in to support meritorious enterprises during the few critical years until they can be either established on a permanent basis or have completed their work.

I recognize of course the weight of certain objections which have been raised, but I think we must run the risk all the same, for my experience teaches me that there is seldom any vitality without antagonism; and the main ground of objection is that we are going on so nicely, we never disagree and therefore we had better remain as we are. But after all, our progress is only obtained by those having differences of opinion coming together and adjusting their differences.

Even should the general opinion be against me, and if it were definitely decided that the International Association of Academies should forever maintain its present state of poverty, the establishment of a domicile on a moderate scale will have to be considered as an independent issue. It might be mentioned that in the original proposals of the Berlin Academy, they intended that there should be not only a central bureau but an organ, published monthly or quarterly, giving an account of the work done by any one academy that would interest the other academies.

The policy which the International Association of Academies will adopt on these questions is one of the most vital importance, for not only will the future of international work depend on the course taken, but the reputation and influence of the academies themselves will, I am convinced, be seriously affected by the decision.

It is with the greatest hesitation and with much diffidence that I now approach the concluding portion of my discourse, for I am oppressed by the fear that my remarks may be taken as an unnecessary interference in the concern of others. But the issue is too serious to let that prevent my expressing an opinion which is based on a deep, and I believe impartial conviction.

The academies, royal societies, or whatever name they are called by, have been founded at different times in accordance with the varying requirements of their countries. They value their historical traditions above everything; some are over two hundred years old, others of recent growth, and their constitutions differ in many respects. But whatever their constitution and their history may be, they must be judged by this same test: Do they fulfill their obligations, which for all of them, I take it, is that defined in the charter of the Royal Society as " The promotion of natural knowledge "? Do they embody in themselves the promotive power of the scientific efforts of their country, or have they fallen a prey to the dangers, which more especially beset the older institutions, of crystalizing into an aristocracy of science, recruited from those who in the natural course of growing maturity are ceasing to be active workers and

constitute themselves to be the judges of the work of others?
The dead weight of such a society, brought to bear discreetly on
the exuberance of youth may have its uses, but it remains a
dead weight just the same. It should act as a brake on a too
fanciful imagination, but it can take no share in any real progress. If the academies are to fit themselves for the formation
of a really strong and fruit bearing association, they must be
bodies which, animated, as all of them now are, by the highest
and noblest ideals, strive at the same time to represent what is
best and most progressive in the scientific life within their range
of influence.

Each country must solve its own difficulties, but in addressing
your National Academy which, though it holds today its first
jubilee, may still be called youthful, I may be forgiven if I
remind you that, while the older institutions may offer you much
that deserves to be admired and perhaps be imitated, you must
not mistake the signs of gray hairs for the stamp of an enviable
dignity.

This, then, is my final summary. Ours is an age of organization presenting many problems that can not be confined within
political boundaries. The demands of science have already
called into existence separate international associations, which
are efficiently performing their duties. Nevertheless the continued increase of their number is beginning to cause incon-
venience and is likely to hamper future developments unless
they can be united by some bond intended to co-ordinate their
work. The International Association of Academies stands out
as a natural body, fit to act as a central advisory authority. To
exercise that authority effectively, the academies must individ-
ually recognize their obligations to be truly representative of
the most healthy and vigorous portion of the scientific life of
their country. It is because I believe in the vitality of the
academies and in the power which an increased responsibility
will give them to check the danger of stagnation to which ancient
and dignified bodies are exposed, that I advocate the extension
of their activity and the more vigorous exercise of the dormant

power which resides in the union of the illustrious bodies which together constitute the International Association of Academies. (Applause.)

THE PRESIDENT: There will be an address this afternoon in this auditorium at 3 o'clock by Dr. Hale on the earth and sun as magnets.

(Whereupon at 12:50 o'clock p. m. a recess was taken until 3 o'clock p. m.)

AFTERNOON SESSION

ADDRESS OF DR. GEORGE ELLERY HALE
Mount Wilson Solar Observatory
ON
THE EARTH AND SUN AS MAGNETS

In 1891, Professor Arthur Schuster, speaking before the Royal Institution, asked a question which has been widely debated in recent years: " Is every large rotating body a magnet? " Since the days of Gilbert, who first recognized that the earth is a great magnet, many theories have been advanced to account for its magnetic properties. Biot, in 1805, ascribed them to a relatively short magnet near its center. Gauss, after an extended mathematical investigation, substituted a large number of small magnets, distributed in an irregular manner, for the single magnet of Biot. Grover suggested that terrestrial magnetism may be caused by electric currents, circulating around the earth and generated by the solar radiation. Soon after Rowland's demonstration in 1876 that a rotating electrically charged body produces a magnetic field, Ayrton and Perry attempted to apply this principle to the case of the earth. Rowland at once pointed out a mistake in their calculation, and showed that the high potential electric charge demanded by their theory could not possibly exist on the earth's surface. It remained for Schuster to suggest that a body made up of molecules which are neutral in the ordinary electrical or magnetic sense may nevertheless develop magnetic properties when rotated.

We shall soon have occasion to examine the two hypotheses advanced in support of this view. While both are promising, it can not be said that either has been sufficiently developed to explain completely the principal phenomena of terrestrial magnetism. If we turn to experiment, we find that iron globes, spun at great velocity in the laboratory, fail to exhibit magnetic properties. But this can be accounted for on either hypothesis. What

4,

we need is a globe of great size, which has been rotating for centuries at high velocity. The sun, with a diameter one hundred times that of the earth (Fig. 1*), may throw some light on the problem. Its high temperature probably precludes the existence of permanent magnets: hence any magnetism it may exhibit is presumably due to motion. Its great mass and rapid linear velocity of rotation should produce a magnetic field much stronger than that of the earth. Finally, the presence in its atmosphere of glowing gases, and the well-known effect of magnetism on light, should enable us to explore its magnetic field even at the distance of the earth. The effects of ionization, probably small in the region of high pressure beneath the photosphere and marked in the solar atmosphere, must be determined and allowed for. But with this important limitation, the sun may be used by the physicist for an experiment which can not be performed in the best equipped laboratory.

Schuster, in the lecture already cited, remarked:

" The form of the corona suggests a further hypothesis which, extravagant as it may appear at present, may yet prove to be true. Is the sun a magnet? "

Summing up the situation in April, 1912, he repeated:

" The evidence (whether the sun is a magnet) rests entirely on the form of certain rays of the corona, which—assuming that they indicate the path of projecting particles—seem to be deflected as they would be in a magnetic field, but this evidence is not at all decisive."

There remained the possibility of an appeal to a conclusive test of magnetism: the characteristic changes it produces in light which originates in a magnetic field.

Before describing how this test has been applied, let us rapidly recapitulate some of the principal facts of terrestrial magnetism. You see upon the screen the image of a steel sphere (Fig. 2), which has been strongly magnetized. If iron filings are sprinkled over the glass plate that supports it, each minute particle becomes a magnet under the influence of the sphere.

* The clichés used in illustration of Dr. Hale's address were courteously furnished by the Editor of *Popular Science Monthly* in which journal the address has been published (August, 1913).—*The Home Secretary.*

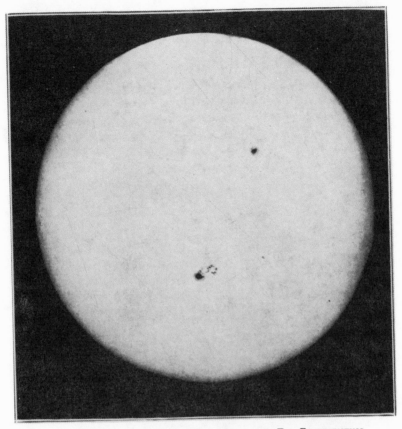

FIG. 1. DIRECT PHOTOGRAPH OF THE SUN WITH DOT REPRESENTING
EARTH FOR COMPARISON.

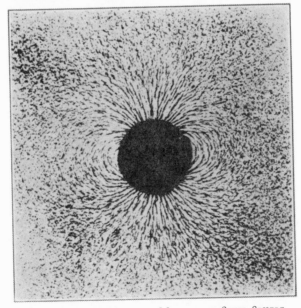

FIG. 2. LINES OF FORCE OF A MAGNETIZED STEEL SPHERE.

When the plate is tapped, to relieve the friction, the particles fall into place along the lines of force, revealing a characteristic pattern of great beauty. A small compass needle, moved about the sphere, always turns so as to point along the lines of force. At the magnetic poles, it points toward the center of the sphere. Midway between them, at the equator, it is parallel to the diameter joining the poles.

As the earth is a magnet, it should exhibit lines of force resembling those of the sphere. If the magnetic poles coincided with the poles of rotation, the north-seeking end of a freely suspended magnetic needle should point vertically downward at one pole, and vertically upward at the other, and the needle should be horizontal at the equator. A dip-needle, used to map the lines of force of the earth, is shown on the screen. I have chosen for illustration an instrument designed for use at sea, on the non-magnetic yacht *Carnegie* (Fig. 3), partly because the equipment used by Dr. Bauer in his extensive surveys represents the best now in use, and also because I wish to contrast the widely different means employed by the Carnegie Institution for the investigation of solar and terrestrial magnetic phenomena. The support of the dip-needle is hung in gimbals, so that observations may be taken when the ship's deck is inclined. The smallest possible amount of metal enters into the construction of this vessel, and where its use could not be avoided, bronze was employed instead of iron or steel. She is thus admirably adapted for magnetic work, as is shown by the observations secured on voyages already totaling more than 100,000 miles. Her work is supplemented by that of land parties, bearing instruments to remote regions where magnetic observations have never before been made.

The dip-needle clearly shows that the earth is a magnet, for it behaves in nearly the same way as the little needle used in our experiment with the magnetized sphere. But the magnetic poles of the earth do not coincide with the geographical poles. The north magnetic pole, discovered by Ross and last visited by Amundsen in 1903, lies near Baffin's Bay, in latitude 70° north, longitude 97° west. The position of the south magnetic pole, calculated from observations made in its vicinity by Captain Scott, of glorious memory, in his expedition of 1901-04, is 72° 50' south latitude, 153° 45' east longitude. Thus the two magnetic poles are not only displaced about 20° from the geographical poles, they do not even lie on the same diameter of the earth. Moreover, they are not fixed in position, but appear to be moving in paths near the geographical poles in a period not yet definitely known. In addition to these peculiarities, it must be added that the dip-needle shows the existence of local magnetic poles, one of which has recently been found by Dr. Bauer at Treadwell Point, Alaska. At such a place the direction of the needle undergoes rapid change as it is moved about the local pole.

The dip-needle, as we have seen, is free to move in a vertical plane. The compass needle moves in a horizontal plane. In general, it tends to point toward the magnetic pole, and as this does not correspond with the geographical pole, there are not many places on the earth's surface where the needle indicates true north and south. Local peculiarities, such as deposits of iron ore, also affect its direction very materially. Thus a variation chart, which indicates the deviation of the compass needle from geographical north, affords an excellent illustration of the irregularities of terrestrial magnetism. The necessity for frequent and accurate surveys of the earth's magnetic field is illustrated by the fact that the *Carnegie* has found errors of five or six degrees in the variation charts of the Pacific and Indian oceans.

In view of the earth's heterogeneous structure, which is sufficiently illustrated by its topographical features, marked deviations from the uniform magnetic properties of a magnetized steel sphere are not at all surprising. The phenomenon of the secular variation is one of the peculiarities toward the solution of which both theory and experiment should be directed.

Passing over other remarkable phenomena of terrestrial magnetism, we come to magnetic storms and auroras, which are almost certainly of solar origin.

FIG. 3.   THE NON-MAGNETIC YACHT CARNEGIE.

FIG. 4.   DIRECT PHOTOGRAPH OF PART OF THE SUN,
APRIL 30, 1908.

Here is a photograph of the sun, as it appears in the telescope (Fig. 4*). Scattered over its surface are sun-spots, which increase and decrease in number in a period of about 11.3 years. It is well known that a curve, showing the number of spots on the sun, is closely similar to a curve representing the variations of intensity of the earth's magnetism. The time of maximum sun-spots corresponds, as Dr. Bauer found, with that of reduced intensity of magnetization of the earth, and the parallelism of the two curves is too close to be the result of accident. We may therefore conclude that there is some connection between the spotted area of the sun and the magnetic field of the earth.

We shall consider a little later the nature of sun-spots, but for the present we may regard them simply as solar storms. When spots are numerous the entire sun is disturbed, and eruptive phenomena, far transcending our most violent volcanic outbursts, are frequently visible. In the atmosphere of the sun, gaseous prominences rise to great heights. This one (shown on the screen), reaching an elevation of 85,000 miles, is of the quiescent type, which changes gradually in form and is abundantly found at all phases of the sun's activity. But such eruptions as the one of March 25, 1895, photographed with the spectroheliograph of the Kenwood Observatory, are clearly of an explosive nature. As these photographs show, it shot upward through a distance of 146,000 miles in 24 minutes, after which it faded away.

When great and rapidly changing spots, usually accompanied by eruptive prominences, are observed on the sun, brilliant displays of the aurora (Fig. 7) and violent magnetic storms are often reported. The magnetic needle, which would record a smooth straight line on the photographic film if it were at rest, trembles and vibrates, drawing a broken and irregular curve. Simultaneously, the aurora flashes and pulsates, sometimes lighting up the sky with the most brilliant display of red and green discharges.

Birkeland and Störmer have worked out a theory which accounts in a very satisfactory way for these phenomena. They

* Figs. 4, 5 and 6 represent the same region of the sun, photographed at successively higher levels.

suppose that electrified particles, shot out from the sun with great velocity, are drawn in toward the earth's magnetic poles along the lines of force. Striking the rarified gases of the upper atmosphere, they illuminate them, just as the electric discharge lights up a vacuum tube. There is reason to believe that the highest part of the earth's atmosphere consists of rarified hydrogen, while nitrogen predominates at a lower level. Some of the electrons from the sun are absorbed in the hydrogen, above a height of 60 miles. Others reach the lower-lying nitrogen, and descend to levels from 30 to 40 miles above the earth's surface. Certain still more penetrating rays sometimes reach an altitude of 25 miles, the lowest hitherto found for the aurora. The passage through the atmosphere of the electrons which cause the aurora also gives rise to the irregular disturbances of the magnetic needle observed during magnetic storms.

The outflow of electrons from the sun never ceases, if we may reason from the fact that the night sky is at all times feebly illuminated by the characteristic light of the aurora. But when sun-spots are numerous, the discharge of electrons is most violent, thus explaining the frequency of brilliant auroras and intense magnetic storms during sun-spot maxima. It should be remarked that the discharge of electrons does not necessarily occur from the spots themselves, but rather from the eruptive regions surrounding them.

Our acquaintance with vacuum tube discharges dates from an early period, but accurate knowledge of these phenomena may be said to begin with the work of Sir William Crookes in 1876. A glass tube, fitted with electrodes, and filled with any gas, is exhausted with a suitable pump until the pressure within it is very low. When a high voltage discharge is passed through the tube, a stream of negatively charged particles is shot out from the cathode, or negative pole, with great velocity. These electrons, bombarding the molecules of the gas within the tube, produce a brilliant illumination, the character of which depends upon the nature of the gas. The rare hydrogen gas in the upper atmosphere of the earth, when bombarded by electrons from the

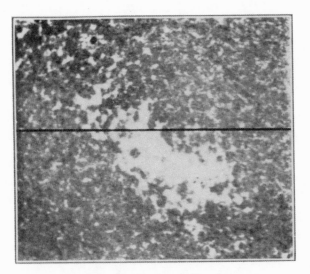

FIG. 5. SAME REGION OF THE SUN, SHOWING
THE CALCIUM (H_2) FLOCCULI.

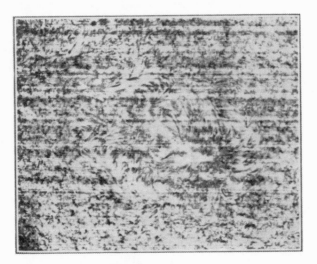

FIG. 6. SAME REGION OF THE SUN, SHOWING
THE HYDROGEN $(H\alpha)$ FLOCCULI.

sun, glows like the hydrogen in this tube. Nitrogen, which is characteristic of a lower level, shines with the light which can be duplicated here.

But it may be remarked that this explanation of the aurora is only hypothetical, in the absence of direct evidence of the emission of electrons by the sun. However, we do know that hot bodies emit electrons. Here is a carbon filament in an exhausted bulb. When heated white hot, a stream of electrons passes off. Falling upon this electrode, the electrons discharge the electroscope with which it is connected. Everyone who has to discard old incandescent lamps is familiar with the result of this outflow. The blackening of the bulbs is due to finely divided carbon carried away by the electrons, and deposited upon the glass.

Now we know that great quantities of carbon in a vaporous state exist in the sun, and that many other substances, also present there, emit electrons in the same way. Hence we may infer that electrons are abundant in the solar atmosphere.

The temperature of the sun is between 6000° and 7000° C., twice as high as we can obtain by artificial means. Under solar conditions, the velocity of the electrons emitted in regions where the pressure is not too great may be sufficient to carry them to the earth. Arrhenius holds that the electrons attach themselves to molecules or groups of molecules, and are then driven to the earth by light-pressure.

In certain regions of the sun, we have strong evidence of the existence of free electrons. This leads us to the question of solar magnetism and suggests a comparison of the very different conditions in the sun and earth. Much alike in chemical composition, these bodies differ principally in size, in density and in temperature. The diameter of the sun is more than one hundred times that of the earth, while its density is only one quarter as great. But the most striking point of difference is the high temperature of the sun, which is much more than sufficient to vaporize all known substances. This presumably means that no permanent magnetism, such as is exhibited by a steel magnet or a lodestone, can exist in the sun. For if we bring this steel magnet to a red heat, it loses its magnetism, and drops the iron bar which it previously supported. Hence, while some theories attribute terrestrial magnetism to the presence within the earth of permanent magnets, no such theory is likely to apply to the sun. If magnetic phenomena are to be found there, they must result from other causes so far as we can judge from our present knowledge of magnetic phenomena at high temperatures.

The familiar case of the helix illustrates how a magnetic field is produced by an electric current flowing through a coil of wire. But according to the modern theory, an electric current is a stream of electrons. Thus a stream of electrons in the sun should give rise to a magnetic field. If the electrons were whirled in a powerful vortex, resembling our tornadoes or water-spouts, the analogy with the wire helix would be exact, and the magnetic field might be sufficiently intense to be detected by spectroscopic observations.

A sun-spot, as seen with a telescope or photographed in the ordinary way, does not appear to be a vortex. If we examine the solar atmosphere above and about the spots, we find extensive clouds of luminous calcium vapor, invisible to the eye, but easily photographed with the spectroheliograph, by admitting no light to the sensitive plate except that radiated by calcium vapor. These calcium flocculi (Fig. 5), like the cumulus clouds of the earth's atmosphere, exhibit no well-defined linear structure. But if we photograph the sun with the red light of hydrogen, we find a very different condition of affairs (Fig. 6). In this higher region of the solar atmosphere, first photographed on Mount Wilson in 1908, cyclonic whirls, centering in sun-spots, are clearly shown.

The idea that sun-spots may be solar tornadoes, which was strongly suggested by such photographs, soon received striking confirmation. A great cloud of hydrogen, which had hung for several days on the edge of one of these vortex structures, was suddenly swept into the spot at a velocity of about 60 miles per second. More recently Slocum has photographed at the Yerkes Observatory a prominence at the edge of the sun, flowing into a spot with a somewhat lower velocity.

FIG. 7. THE AURORA.

FIG. 8. WATER-SPOUT.

Thus we were led to the hypothesis that sun-spots are closely analogous to tornadoes or water-spouts in the earth's atmosphere (Fig. 8). If this were true, electrons, caught and whirled in the spot vortex, should produce a magnetic field. Fortunately, this could be put to a conclusive test, through the well-known influence of magnetism on light discovered by Zeeman in 1896.

In Zeeman's experiment a flame containing sodium vapor was placed between the poles of a powerful electro-magnet. The two yellow sodium lines, observed with a spectroscope of high dispersion, were seen to widen the instant a magnetic field was produced by passing a current through the coils of the magnet. It was subsequently found that most of the lines of the spectrum, which are single under ordinary conditions, are split into three components when the radiating source is in a sufficiently intense magnetic field. This is the case when the observation is made at right angles to the lines of force. When looking along the lines of force, the central line of such a triplet disappears (Fig. 9), and the light of the two side components is found to be circularly polarized in opposite directions. With suitable polarizing apparatus, either component of such a line can be cut off at will, leaving the other unchanged. Furthermore, a double line having these characteristic properties can be produced only by a magnetic field. Thus it becomes a simple matter to detect a magnetic field, at any distance, by observing its effect on light emitted within the field. If a sun-spot is an electric vortex, and the observer is supposed to look along the axis of the whirling vapor, which would correspond with the direction of the lines of force, he should find the spectrum lines double, and be able to cut off either component with the polarizing attachment of his spectroscope.

I applied this test to sun-spots on Mount Wilson in June, 1908, with the 60-foot tower telescope, and at once found all of the characteristic features of the Zeeman effect. Most of the lines of the sun-spot spectrum are merely widened by the magnetic field, but others are split into separate components (Fig. 10), which can be cut off at will by the observer. Moreover, the opportune formation of two large spots, which appeared on the spectroheliograph plates to be rotating in opposite directions (Fig. 11), permitted a still more exacting experiment to be tried. In the laboratory, where the polarizing apparatus is so adjusted as to transmit one component of a line doubled by a magnetic field, this disappears and is replaced by the other component when the direction of the current is reversed. In other words, one component is visible alone when the observer looks toward the north pole of the magnet, while the other appears alone when he looks toward the south pole. If electrons of the same kind are rotating in opposite directions in two sun-spot vortices, the observer should be looking toward a north pole in one spot and toward a south pole in the other. Hence the opposite components of a magnetic double line should appear in two such spots. As our photographs show, the result of the test was in harmony with my anticipation.

I may not pause to describe the later developments of this investigation, though two or three points must be mentioned. The intensity of the magnetic field in sun-spots is sometimes as high as 4500 gausses, or nine thousand times the intensity of the earth's field. In passing upward from the sun's surface, the magnetic intensity decreases very rapidly—so rapidly, in fact, as to suggest the existence of an opposing field. It is probable that the vortex which produces the observed field is not the one that appears on our photograph, but lies at a lower level. In fact, the vortex structure shown on spectroheliograph plates may represent the effect, rather than the cause of the sun-spot field. We may have, as Brester and Deslandres suggest, a condition analogous to that illustrated in the aurora: electrons, falling in the solar atmosphere, move along the lines of force of the magnetic field into spots. In this way we may perhaps account for the structure surrounding pairs of spots, of opposite polarity, which constitute the typical sun-spot group. The resemblance of the structure near these two bipolar groups to the lines of force about a bar magnet is very striking, especially when the disturbed condition of the solar atmosphere, which tends to mask the effect, is borne in

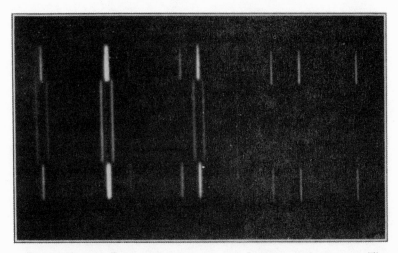

FIG. 9. ZEEMAN DOUBLET PHOTOGRAPHED IN LABORATORY SPECTRUM. The middle section shows the doublet. The adjacent sections indicate the appearance of the spectrum line in the absence of a magnetic field.

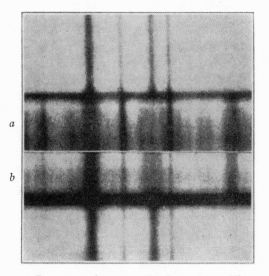

FIG. 10. *a, b,* spectra of two sun-spots. The triple line indicates a magnetic field of 4500 gausses in *a*, and one of 2900 gausses in *b*.

5

mind. It is not unlikely that the bipolar group is due to a single vortex, of the horse-shoe type, such as we may see in water after every sweep of an oar.

We thus have abundant evidence of the existence on the sun of local magnetic fields of great intensity—fields so extensive that the earth is small in comparison with many of them. But how may we account for the copious supply of electrons needed to generate the powerful currents required in such enormous electro-magnets? Neutral molecules, postulated in theories of the earth's field, probably will not suffice. A marked preponderance of electrons of one sign seems to be indicated.

An interesting experiment, due to Harker, will help us here. Imagine a pair of carbon rods, insulated within a furnace heated to a temperature of two or three thousand degrees. The outer ends of the rods, projecting from the furnace, are connected to a galvanometer. Harker found that when one of the carbon terminals within the furnace was cooler than the other, a stream of negative electrons flowed toward it from the hotter electrode. Even at atmospheric pressure, currents of several amperes were produced in this way.*

Our spectroscopic investigations, interpreted by laboratory experiments, are in harmony with those of Fowler in proving that sun-spots are comparatively cool regions in the solar atmosphere. They are hot enough, it is true, to volatilize such refractory elements as titanium, but cool enough to permit the formation of certain compounds not found elsewhere in the sun. Hence, from Harker's experiment, we may expect a flow of negative electrons toward spots. These, caught and whirled in the vortex, would easily account for the observed magnetic fields.

The conditions existing in sun-spots are thus without any close parallel among the natural phenomena of the earth. The sun-spot vortex is not unlike a terrestrial tornado, on a vast scale, but if the whirl of ions in a tornado produces a magnetic field, it is too feeble to be readily detected. Thus, while we have dem-

---

* King has recently found that the current decreases very rapidly as the pressure increases, but is still appreciable at a pressure of 20 atmospheres.

onstrated the existence of solar magnetism, it is confined to limited areas. We must look further if we would throw new light on the theory of the magnetic properties of rotating bodies.

This leads us to the question with which we started: is the sun a magnet like the earth? The structure of the corona, as revealed at total eclipses, points strongly in this direction. Remembering the lines of force of our magnetized steel sphere, we can not fail to be struck by their close resemblance to the polar streamers in these beautiful photographs of the corona (Fig. 12) taken by Lick Observatory eclipse parties, for which I am indebted to Professor Campbell. Bigelow, in 1889, investigated this coronal structure, and showed that it is very similar to the lines of force of a spherical magnet. Störmer, guided by his own researches on the aurora, has calculated the trajectories of electrons moving out from the sun under the influence of a general magnetic field, and compared these trajectories with the coronal streamers.* The resemblance is apparently too close to be the result of chance. Finally, Deslandres has investigated the forms and motions of solar prominences, which he finds to behave as they would in a magnetic field of intensity about one millionth that of the earth. We may thus infer the existence of a general solar magnetic field. But since the sign of the charge of the outflowing electrons is not certainly known, we can not determine the polarity of the sun in this way. Furthermore, our present uncertainty as to the proportion at different levels of positive and negative electrons, and of the perturbations due to currents in the solar atmosphere, must delay the most effective application of these methods, though they promise much future knowledge of the magnetic field at high levels in the solar atmosphere.

Of the field at low levels, however, they may tell us little or nothing. To detect this low-level field we must resort to the method employed in the case of sun-spots—the study of the Zeeman effect. If this is successful, it will not only show beyond doubt whether the sun is a magnet: it will also permit the polarity

* See also the important investigations of Birkeland.

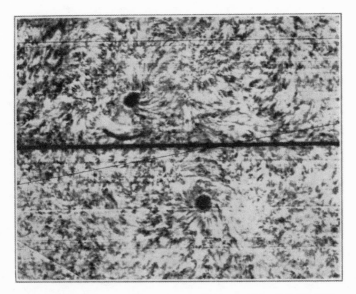

FIG. 11. RIGHT- AND LEFT-HANDED VORTICES SURROUNDING SUN-
SPOTS, as indicated by the distribution of hydrogen gas. Photographed with the Spectroheliograph.

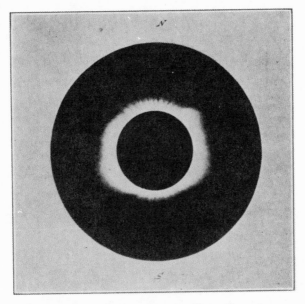

FIG. 12. SOLAR CORONA, SHOWING POLAR STREAMERS.

of the sun to be compared with that of the earth, give a measure of the strength of the field at different latitudes, and indicate the sign of the charge that a rotating sphere must possess if it is to produce a similar field.

I first endeavored to apply this test with the 60-foot tower telescope in 1908, but the results were too uncertain to command confidence.

Thanks to additional appropriations from the Carnegie Institution of Washington, a new and powerful instrument was available on Mount Wilson for a continuation of the investigation in January, 1912. The new tower telescope has a focal length of 150 feet (Fig. 13). To prevent vibration in the wind, the cœlostat, second mirror and object-glass are carried by a skeleton tower, each vertical and diagonal member of which is enclosed within the corresponding member of an outer skeleton tower, which also carries a dome to shield the instruments from the weather. In the photograph, we see only the hollow members of the outer tower. But within each of them, well separated from possible contact, a sectional view would show the similar, but more slender members of the tower that supports the instruments. The plan has proved to be successful, permitting observations demanding the greatest steadiness of the solar image to be made.

The arrangements are similar to those of the 60-foot tower. The solar image, $16\frac{1}{2}$ inches in diameter, falls on the slit of a spectrograph (Fig. 14) in the observation house at the ground level. The spectrograph, of 75 feet focal length, enjoys the advantage of great stability and constancy of temperature in its subterranean vault beneath the tower. In the third order spectrum, used for this investigation, the D lines of the solar spectrum are 29 millimeters apart. The resolving power of the excellent Michelson grating is sufficient to show 75 lines of the iodine absorption spectrum in this space between the D's. Thus the instruments are well suited for the exacting requirements of a difficult investigation. For it must be borne in mind that the problem is very different from that of detecting the magnetic fields in sun-spots, where the separation of the lines is from fifty to one hundred times as great as we may expect to find here. Thus the sun's general field can produce no actual separation of the lines. But it may cause a very slight widening, which should appear as a displacement when suitable polarizing apparatus is used. This is so arranged as to divide the spectrum longitudinally into narrow strips. The component toward the red end of the spectrum of a line widened by magnetism should appear in one strip, the other component in the next strip. Hence, if the sun has a magnetic field of sufficient strength, the line should have a dentated appearance. The small relative displacements of the lines on successive strips, when measured under a microscope, should give the strength of the magnetic field.

The above remarks apply strictly to the case when the observer is looking directly along the lines of force. At other angles neither component is completely cut off, and the magnitude of the displacement will then depend upon two things: the strength of the magnetic field and the angle between the line of sight and the lines of force. Assuming that the lines of force of the sun correspond with those of a magnetized sphere, and also that the magnetic poles coincide with the poles of rotation, it is possible to calculate what the relative displacement should be at different solar latitudes. These theoretical displacements are shown graphically by the sine curve on the screen (Fig. 15).

We see from the curve that the greatest displacements should be found at 45° north and south latitude, and that from these points they should decrease toward zero at the equator and the poles. Furthermore, the curve shows that we may apply the same crucial test used in the case of sun-spots: the direction of the displacements, toward red or violet, should be reversed in the northern and southern hemispheres.

I shall not trouble you with the details of the hundreds of photographs and the thousands of measures which have been made by my colleagues and myself during the past year. In view of the diffuse character of the solar lines under such high dispersion, and the exceedingly small displacements observed,

FIG. 13. 150-FOOT TOWER TELESCOPE.

the results must be given with some reserve, though they appear to leave no doubt as to the reality of the effect. Observations in the second order spectrum failed to give satisfactory indications of the field. But with the higher dispersion of the third order, eleven independent determinations, made with every possible precaution to eliminate bias, show opposite displacements in the northern and southern hemispheres, decreasing in magnitude from about 45° north and south latitude to the equator. Three of these determinations were pushed as close to the poles as conditions would permit, and the observed displacements may be compared with the theoretical curve (Fig. 15). In view of the very small magnitude of the displacements, which never surpass 0.002 Angströms, the agreement is quite as satisfactory as one could expect for a first approximation.

The full details of the investigation are given in a paper recently published.* The reader will find an account of the precautions taken to eliminate error, and, I trust, no tendency to underestimate the possible adverse bearing of certain negative results. It must remain for the future to confirm or to overthrow the apparently strong evidence in favor of the existence of a true Zeeman effect, due to the general magnetic field of the sun. If this evidence can be accepted, we may draw certain conclusions of present interest.

Taking the measures at their face value, they indicate that the north magnetic pole of the sun lies at or near the north pole of rotation, while the south magnetic pole lies at or near the south pole of rotation. In other words, if a compass needle could withstand the solar temperature, it would point approximately as it does on the earth, since the polarity of the two bodies appears to be the same. Thus, since the earth and sun rotate in the same direction, the preponderant influence of a negative charge distributed through their mass (in combination with an equal positive charge of slightly different volume density) would account in each case for the observed magnetic polarity.

* *Contributions from the Mount Wilson Solar Observatory,* No. 71.

As for the strength of the sun's field, only three preliminary determinations have yet been made, with as many different lines. Disregarding the systematic error of measurement, which is still very uncertain, these indicate that the field-strength at the sun's poles is of the order of fifty gausses (about eighty times that of the earth).

Schuster, assuming the magnetic fields of the earth and sun to be due to their rotation, found that the strength of the sun's field should be 440 times that of the earth, or 264 gausses.* This was on the supposition that the field-strength of a rotating body is proportional to the product of the radius and the maximum linear velocity of rotation, but neglected the density. Before inquiring why the observed and theoretical values differ, we may glance at the two most promising hypotheses that have been advanced in support of the view that every large rotating body is a magnet.

On account of their greater mass, the positive electrons of the neutral molecules within the earth may perhaps be more powerfully attracted by gravitation than the negative electrons. In this case the negative charge of each molecule should be a little farther from the center of the earth than the positive charge. The average linear velocity of the negative charge would thus be a little greater, and the magnetizing effect due to its motion would slightly exceed that due to the motion of the positive charge. By assuming a separation of the charges equal to about four-tenths the radius of a molecule (Bauer), the symmetrical part of the earth's magnetic field could be accounted for as the result of the axial rotation.†

This theory, first suggested by Thomson, has been developed by Sutherland, Schuster and Bauer. But as yet it has yielded no explanation of the secular variation of the earth's magnetism, and other important objections have been urged against it.

* Bauer, by a similar method, obtained 306 gausses.

† According to Bauer's view, in each molecule one charge occupies a larger volume than the other, so as to make the volume densities of the two charges at the same point slightly different.

FIG. 14. HEAD OF THE 75-FOOT SPECTOGRAPH OF THE 150-FOOT TOWER TELESCOPE.

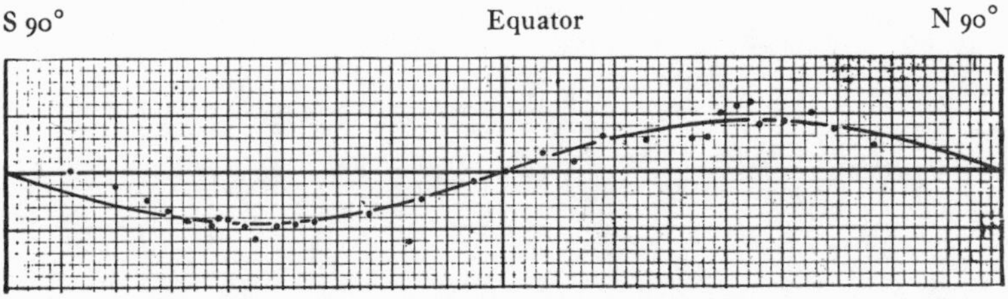

S 90°   Equator   N 90°

FIG. 15. The curve represents the theoretical variation of the displacements of spectrum lines with the heliographic latitude. The sun is assumed to be a magnetized sphere with its magnetic poles coinciding with the poles of rotation. The points represent mean values of the observed displacements. Vertical scale: 1 square = 0.001 mm. = 0.0002 Ångström.

While it should not be rejected, the merits of other theories must not be overlooked.

Chief among these is the theory that rests on the very probable assumption that every molecule is a magnet. If the magnetism is accounted for as the effect of the rapid revolution of electrons within the molecule, a gyrostatic action might be anticipated. That is, each molecule would tend to set itself with its axis parallel to the axis of the earth, just as the gyrostatic compass, now coming into use at sea, tends to point to the geographical pole. The host of molecular magnets, all acting together, might account for the earth's magnetic field.

This theory, in its turn, is not free from some points of weakness, though they may disappear as the result of more extended investigation. Its chief advantage lies in the possibility that it may perhaps explain the secular variation of the earth's magnetism by a precessional motion of the magnetic molecules.

On either hypothesis, it is assumed, in the absence of knowledge to the contrary, that every molecule contributes to the production of the magnetic field. Thus the density of the rotating body may prove to be a factor. Perhaps the change of density from the surface to the center of the sun must also be taken into account. But the observational results already obtained suggest that the phenomena of ionization in the solar atmosphere may turn out to be the predominant influence.

The lines which show the Zeeman effect originate at a comparatively low level in the solar atmosphere. Preliminary measures indicate that certain lines of titanium, which are widely separated by a magnetic field in the laboratory, are not appreciably affected in the sun. As these lines represent a somewhat higher level, it is probable that the strength of the sun's field decreases very rapidly in passing upward from the surface of the photosphere—a conclusion in harmony with results obtained from the study of the corona and prominences. Thus it may be found that the distribution of the electrons is such as to give rise to the observed field or to produce a field opposing that caused by the rotation of the body of the sun. It is evident that speculation along these lines may advantageously await the accumulation of observations covering a wide range of level. Beneath the photosphere, where the pressure is high, we may conclude from recent electric furnace experiments by King that free electrons, though relatively few, may nevertheless play some part in the production of the general magnetic field.

In this survey of magnetic phenomena, we have kept constantly in mind the hypothesis that the magnetism of the earth is due to its rotation. Permanent magnets, formerly supposed to account for the earth's magnetic field, probably can not exist at the high temperature of the sun. Displays of the aurora, usually accompanied by magnetic storms, are plausibly attributed to electrons reaching the earth from the sun, and illuminating the rare gases of the upper atmosphere just as they affect those in a vacuum tube. Definite proof of the existence of free electrons in the sun is afforded by the discovery of powerful local magnetic fields in sun-spots, where the intensity of the field is sometimes as great as nine thousand times that of the average value of the earth's field. These local fields probably result from the rapid revolution in a vortex of negative electrons, flowing toward the cooler spot from the hotter region outside. The same method of observation now indicates that the whole sun is a magnet, of the same polarity as the earth. Because of the high solar temperature, this magnetism may be ascribed to the sun's axial rotation.* It is not improbable that the earth's magnetism also results from its rotation, and that other rotating celestial bodies, such as stars and nebulæ, may ultimately be found to possess magnetic properties. Thus, while the presence of free electrons in the sun prevents our acceptance of the evidence as a proof that every large rotating body is a magnet, the results of the investigation are not opposed to this hypothesis, which may be tested further by the study of other stars of known diameter and velocity of rotation.

---

* The alternative hypothesis, that the sun's magnetism is due to the combined effect of numberless local magnetic fields, caused by electric vortices in the solar " pores," though at first sight improbable, deserves further consideration.

SECOND DAY

SMITHSONIAN INSTITUTION, WASHINGTON, D. C.

Wednesday, April 23, 1913.

The meeting was called to order by President Remsen at 10: 40 o'clock a. m.

THE PRESIDENT: I regret to have to announce that Professor Boveri is ill and will be unable to give the promised lecture.

The next speaker is the distinguished astronomer of Holland, Dr. J. C. Kapteyn, who is Director of the Astronomical Laboratory. I was told that, unless I stopped there for a moment and emphasized that word, it would be assumed that I had made a mistake or that somebody else had made a mistake; but that is the name of the institution over which Dr. Kapteyn presides. It is not an astronomical observatory. Those two names seem to flow together by a natural process; but it is an astronomical laboratory, where work proper to a laboratory is carried on, and where work proper to an astronomical observatory can not be carried on.

I have pleasure in presenting to you Dr. Kapteyn, Director of the Astronomical Laboratory of the University of Groningen, who will speak to you on " The Structure of the Universe." Dr. Kapteyn. (Applause.)

ADDRESS OF DOCTOR J. C. KAPTEYN
ON
THE STRUCTURE OF THE UNIVERSE

DR. KAPTEYN: *Ladies and Gentlemen:* I have been asked to address you on the structure of the universe. The title is ambitious, and I fear that what I have to say on the subject will be sadly in disproportion with what some of you will be led to expect by this title.

It will, however, I hope, give you a glimpse of what astronomers nowadays are attempting to do, in order to penetrate somewhat into the mystery of the starry sky.

The problem, as I take it, is a double one. We have, first, the structure of the universe as it is at the present moment; and this problem is, in the main, no other than finding the star distances, because the star directions we can readily ascertain.

We have, second, the problem of the history and evolution of the system.

As the time at my disposal is so short, I must confine myself to one of the two, and undoubtedly the first is at the present moment the more promising, owing to the recent discovery of star-streaming, and due furthermore to the history of the system during the past ages, ages to be counted by millions, and probably hundreds of millions of years, which is and perhaps will remain enshrouded in great mystery. Still I think that the second problem of the evolution of the system may, perhaps, be the more suitable subject for the present lecture.

You will all, of course, understand, without my saying anything to the purpose, that what we have to expect can not well be anything else than a few more or less probable inferences about the course of events that have made our system what it is.

Some additional considerations might easily have been added, but as I have had to give up the idea of giving a general review of what has been done, I thought it might be as well to confine myself to just a few illustrations of the kind of speculations that we are at present being led to; and as these speculations, mainly or wholly, also depend on the theory of star-streaming, it may be well to begin by saying a few words about that theory.

In order to get a clear idea of what is understood by the phenomenon of star-streaming: Imagine two clouds or swarms of stars, at first wide apart in space; imagine that the stars within each cloud move in all directions, indiscriminately, pretty much as do the molecules of gas, and let us call this motion in the cloud the " internal motion." In fact, imagine two immense gas bubbles, the molecules of which will be our stars.

Now, imagine these two clouds to be moving in space, and let that motion bring the two gas bubbles together, so that they will penetrate each other. Then imagine that we, the spectators, are in that part of the universe where the two bubbles have inter-mixed, and finally imagine that we, the spectators, have a motion of our own.

What we will see of the motion of the individual gas mole-cules will very nearly correspond to what we see of the motion of the stars actually going on in the sky.

Now, how does this motion show? Had the molecules in each gas bubble no internal motion, that is, had they no other motion than the common motion of all the molecules together as a whole gas bubble, then of course what you would see would be this: We would see two immense streams of stars, all moving in per-fectly parallel lines, with perfectly equal velocity. If, however, the internal motion is not zero, then of course what we will see will be more or less different. The internal motion gives to each molecule, in addition to the motion which is common to the whole of the bubble, an additional individual motion, which will be the cause of the total motion of the several molecules that are not perfectly parallel or perfectly equal; and instead of seeing two perfect streams with perfectly parallel motions, we must now see the stars in the main parallel to two great directions; but there will be deviations—small deviations will be frequent, greater deviations will be rare, and very great deviations will be decidedly exceptional. The motions of the two individual bubbles will still be clearly discernible.

So it is that with the stars in the sky we have recognized two clearly defined preferential motions. These directions make an angle of about one hundred degrees. The stars are not moving in these directions only. Small deviations are frequent; greater deviations are somewhat rare; very great deviations are decidedly exceptional. This describes what is actually observed in the sky.

We may say that all investigations made since the first announcement of star-streaming in 1904—investigations based on very different materials—all agree in the establishment of these two preferential directions of motion among the stars. We find them in the brightest stars; we find them in the fainter stars; they show in the swift moving stars; they show in the slow moving stars. They betray their existence not only in the radial motions but as well in the motion at right angles to the visible ray.

Our conception of two independent star clouds is one of these investigations. Whether this interpretation is the correct one, is a question of evolution of the system and will have to be considered presently. Our conclusion will eventually be in favor of the two-cloud theory; and so, for the sake of greater clearness, I will continue to use this conception.

In the study of the history of the system, we start from what we know, or think we know, about the evolution of the separate stars.

The stars have been classified by Sechi into four spectral classes. We have a far more elaborate classification, but for the present purpose Sechi's classification will do. The stars of the fourth type are so few in number that we may, for the present, neglect them.

On the other hand, there is a class of stars showing the helium lines, that formerly was classified with all the other classified stars, but it is now considered as a separate class—the helium stars.

So that now, neglecting the stars of Sechi's fourth type, we have still four types, which will be the helium stars, and the first, second and third types of stars—helium, first, second and third. Now, there is much evidence to show that this classification in this order is a natural one. I mean that the order is really an order of evolution, the helium stars being the stars of recent birth; while we get to the older and older stars as we pass from the helium stars to the first, from the first to the second, and from the second to the third. I will adopt this classification in what follows, although well aware of the fact that all astronomers do not agree with me. I feel justified in this course, not only because I think it is the opinion of the great majority of our eminent spectroscopists, but also because the very facts which I wish to put before you about star-streaming strongly confirm it.

When we wish to penetrate into the history of the system, it seems natural to investigate the problem of star-streaming separately for these four classes of stars in the order of their evolution. There are some difficulties, mainly the consequence of the scantiness of materials. Still, however, even now it has been possible to carry that investigation through in such a way as to establish a few facts, and to give clear indications of other facts. Of these I will consider only the two following, about the reality of which I think there can be left hardly any doubt.

First, the older the stars, the greater the internal velocity, and second, the older the stars, the richer the second stream, at least in comparison with the first stream. And I wish to consider some of the inferences to which these facts lead us. In the first place, the facts at once lead us back to the question just now mentioned, about the order of evolution of the individual stars. For this regularity in the increase in the internal velocity and also in the richness of the second stream exists only if we arrange the stars in their order, helium, first, second and third, or (of course I need not say so) in the reverse order, third, second, first, helium; but in no other arrangement.

Therefore, with the same right that we expect that all the properties of the stars will change with age, gradually, and not per saltum, with that same right, I think, we conclude that the order of evolution must be helium, first, second, third, or the exact reverse.

Now, then, there is much evidence that it is not the reverse order; but I think it is better not to lose time in going into that point just now; and, therefore, I will say that we have strong confirmation here of what, on two or three different grounds, is pretty generally considered as the order of the different ages in a star's life.

But to proceed: As the younger the stars, the smaller their internal motion, it follows at once that if we go to a stage younger still, if we go back to that matter from which the helium stars have been evolved (and I call that matter primordial matter), that internal motion of primordial matter must be smaller still.

And, according to our second fact, the richness of the second stream is smaller and smaller, the younger the class of stars we consider.

The number of the helium stars is already so great that, not long ago, it was assumed that there was no trace of a second stream in the helium stars at all. We conclude, that in this still younger stage of evolution it is practically certain that there can not be a second stream at all.

Therefore, finally, we must expect that the particles of primordial matter will all move in practically parallel lines, in the direction and with the velocity of that stream in which all but the totality of the helium stars move.

It is a very general notion that that primordial matter, that matter from which the stars are evolved, is the matter of the nebulæ. What precedes gives us the means of testing this theory by observation.

What, then, does observation show about the nebulæ? The number of available data is as yet extremely small. The determination of what we call astronomical proper motion of these very ill-defined nebulæ is extremely difficult, and has been up to the present time invariably unsuccessful. We can say that we know nothing about the astronomical motion of the nebulæ. To the determination of the radial velocity with the spectroscope the faintness of the object is a great obstacle. The consequence is that, as yet, we know the radial velocity of only fourteen of these objects in all. Still, this limited number is decisive in showing that there can be no question that the real motions of these objects are not even approximately parallel to the motion of the helium stars, or even parallel to any fixed direction whatever. Their velocity, moreover, is exceedingly unequal. Therefore, must we conclude that the nebulæ are not the birthplace of the stars? It may seem so.

Meanwhile let us not go too far. There are nebulæ and nebulæ, and it so happens—and there is ample practical reason for it—that with one exception observation of radial velocity has, up to the present time, been confined to what we call the planetary nebulæ—nebulæ elliptical or round, which show an appearance remotely like that of a planetary disc. Herschel saw in them a likeness to what, according to Laplace's cosmogony, must have been the primitive stage of our own planetary system; and therefore he thought that these planetary nebulæ must be the birthplace of the stars.

Now then, according to what precedes, this view seems now untenable. If they were, they would have shown the parallel and equal motion of practically all the helium stars. The motions on the contrary, are extremely unparallel, and unequal, and we must assign these objects a place at the end of the order of evolution rather than at the beginning.

We may, perhaps, see an independent confirmation of this view in the stars called temporary stars, but time will not permit me to pursue that argument further.

Meanwhile, as I said just now, there is one exception in which the radial velocity of the nebulæ has been determined, which is not planetary. This exception is the Orion nebula, the well known Orion nebula, which is classified under the irregular nebulæ. Therefore, we might ask if its position with the irregular nebulæ was the birthplace of the stars.

It turns out that this one object has exactly the radial velocity of the helium stars in the first stream; that is, we find exactly the motion we must expect in these nebulæ, if it were the birthplace of the stars. We will not, of course, on this single fact find far-reaching conclusions; but we have a right, in my opinion, to say that here is a fact that singularly strengthens what has already been concluded from other facts.

We see, moreover, that the observation of the radial velocity of other irregular nebulæ must, ere long, furnish us with a crucial test of the theory.

There is another problem involved in our observations which might seem to be of no less importance than the one just mentioned. How have we to explain the fact that the internal velocity of the stars gradually increases with age? The astronomer who, in the study of the motion of the heavenly bodies, has found hardly a trace of any other force than gravitation, will naturally turn to gravitation for such an explanation, and it really seems a necessity, that under the influence of the mutual gravitation, bodies which at the outset have little or no relative motion must get such motion, and they must come to fall toward each other, and this velocity must increase with time.

Thus far, there is no great difficulty, but now let us look farther back in time, back to the time in which the stars had not yet been formed, and in which matter was still in its primordial state. If it be true that mutual attraction of the stars has generated such an enormous amount of internal motion in the time needed by the helium stars to develop from the helium stars to second and third class stars, how have we to explain that we find that same matter nearly at rest at the first stage of evolution at which we meet it? How have we to explain that in pre-helium ages gravitation has produced no effect?

He that believes in the creation of matter at a finitely remote epoch may find no difficulty in the question; but to him who does not, it is simply astonishing to see matter behaving as if there were no gravitation at all. What may be the explanation? Is there no gravitation in primordial matter, or is there another force exactly counterbalancing its effects?

I shall offer no solution. I simply wish to point out that here is a problem which may be interesting to the physicist no less than to the astronomer.

Passing to other inferences, I wish to draw your attention to a question already alluded to. Does the observed fact of the preference of the star motions for two definite directions lead us with necessity to the assumption that our system has been formed by the meeting of two independent star clouds? Or is it still possible, and in that case more plausible, to explain it without the sacrifice of the unity of the system? In other words, is it a dual system, or is it one unit?

Suppose a very elongated system of stars which are originally at rest; now let these be left to their mutual attraction; then it is evident that those stars, under the influence of that gravitation will fall together. The stars at the extreme ends of this elongated system will fall toward each other, in a motion that will follow, in the main, the axis of that system of stars.

In the main, there will be deviations because, of course, the system is pretty long in comparison to its breadth; but still, it is not linear. In other words, there will be set up two streams of stars, in which all the stars will be moving preferentially and there will be deviations. Those star-streams will be exactly opposite each other. There is no difficulty in that. True it is that the streams of stars we see make an angle of about one hundred degrees, but it is evident that if we have two streams moving in exactly opposite directions, and we view these streams from the earth which itself has a motion in space, the consequence will then be that those opposite streams will be seen to make an angle, and it is easy to determine the earth's motion in such a way as to bring it in perfect harmony with the observation. Thus far there is no objection. It might seem that this evolution of the system, which saves the unity of the system, might explain the matter; but there are further consequences.

In an elongated universe, as here supposed, both the longitudinal motion (what we in this lecture have called the stream motion) and the deviations therefrom (what in this lecture have been called the internal motions) must gradually increase from zero to the value we find them having at the present time.

Now as to the internal velocities, this is exactly what we find by observation. Do we find the same for the stream motion?

Recent Mt. Wilson observations have enabled us to derive at least a pretty reliable value of the relative stream velocities for the first type stars. For the helium stars we can as yet only assign a limit which the relative velocities of the two streams must exceed; but for the older stars we have had reliable information for some time.

All these determinations show, contrary to what takes place with the internal motion, that the relative velocity of the stream motion does not change, or does not change very much, at least, with age; certainly the stream motion is not nearly vanishing for the helium stars. It seems to me that this consideration is fatal to the present explanation.

Professor Schwarzschild has developed a different theory, which also leaves the universe a unit; but this theory also can not, I think, be maintained. Among other things, we have, as a main objection, the fact—which was not known at the time Professor Schwarzschild proposed his theory—that the richness of the two streams is not the same for stars of different age.

The tacit assumption is made, and must be made, under Schwarzschild's theory that the two streams have the same number of stars. Now, this may be more or less approximately true of the stars of the second and third types. It certainly is not true of stars of the first type.

While it is true that the first stream is three times richer than the second stream, the richness of the helium stars is certainly not a tenth of that of the first stream.

The conclusion to be drawn from all of this seems pretty obvious. It would seem that we are driven to the theory assumed here, from the first, the theory of the two star-clouds, which, owing to their initial velocity, have come to meet and intermingle in space.

It must be confessed, however, that in this theory also there remain some hard nuts to crack; and until we succeed in this it seems unsafe to claim any great certainty for the theory, and it seems preferable to put it forward as one hypothesis, which, for the time being, best fits the observed facts.

There remains to be considered the question how to explain that the second stream or cloud hardly contains any helium stars. There is something in the small local groups which we know in this type which may help us. Everybody knows the group of the Pleiades.

There can be no doubt but that the bright stars and many of the faint stars that we see in this part of the sky, are near together in space and not merely near together in the same visual line—the one far behind the other. They undoubtedly form a physical system, and must have had a common origin. At present we know several of such local groups, among them the Hyades, the Ursa Major group, and we may perhaps add the great Scorpius Centaur group.

Now, in these local groups, we find, amongst others, two very remarkable facts. The first is that if the stars of such groups are arranged in the order of their brightness, we find that they are arranged, with a few, and it must be confessed, significant exceptions, in the order of the spectral classes. As an instance, take the Scorpius Centaur group. We find that the very brightest stars are of the earliest helium type; the somewhat fainter ones are of the next helium type; the next fainter ones are of the next stage in the stellar life, or the first type; and if we can not follow the series further on to the second and perhaps the third type, this is probably due to our ignorance about it or lack of knowledge of the fainter stars belonging to the group.

In the Pleiades, where we have a somewhat more extensive knowledge of the fainter stars, we can follow the series at least as far as the middle of the second type stars. It follows from this that in all these groups, what there are of helium stars can not be overlooked, for they all are of the very brightest stars, and our knowledge of the brightest stars is pretty complete.

Notwithstanding this, and this is the second remarkable fact, the fact that bears directly on the question in hand—we find not a single helium star, neither in the Hyades nor in the Ursa Major group. The stars in this group show a gradual change of spectrum with the brightness, but they certainly begin with the second stage of the star's life. In the Pleiades, this is somewhat different. Here there are some helium stars. But they begin abruptly with the middle stage of helium stars. There is not a single star of the early helium type in the Pleiades. So that the series begins, not at the beginning, but at some different point in the evolution.

As to our second stream, then, we have to explain that there is no second stream, or no appreciable number of helium stars in the second stream. Our second stream, then, in which we meet with no helium stars is absolutely in the same condition as the group of the Hyades and the Ursa Major group; and the explanation of the two cases must, in all likelihood, be the same.

How, therefore, does it come to pass that in such groups as those of the Hyades and the Ursa Major, the helium stars are absolutely wanting?

For those who, as I did in this lecture, adopt the view of the order of evolution as helium, first, second, third, there can be no question but that the stars which we now see are first type stars, which must in past ages have been helium stars.

Therefore, such a group as the Hyades, which nowadays does not contain any helium stars, but which contains first type stars, must in past ages have contained the helium stars in great numbers; and going back to times somewhat further still, these helium stars must have been evolved from some primordial matter. This primordial matter is possibly some nebulous matter, and therefore, in the still earlier ages, these groups of the Hyades and Ursa Major must have been filed with nebulæ.

Now, then, in the present day, so far as we know, there is no trace of nebulæ in these groups. Our conclusion, therefore, is that there must have been some time in which the nebula in those groups was exhausted. It was probably taken up in the formation of stars. Since the exhaustion of the nebulous matter, evidently there could be formed no more helium stars; and the helium stars that had been formed developed gradually in the second stage into first type stars, and in a short time there will be an end to helium stars in those groups, and this will explain perfectly how it is coming to pass at present, that we do not find any helium stars in those groups.

Therefore, finally, our answer to the question: how does it come to pass that in the second stream or cloud we find hardly any helium stars, would be because at some time the nebulous matter must have been exhausted in this cloud.

As to the first stream or star cloud, we similarly conclude that the nebulous matter must not yet have been exhausted, or, if so, only at a very recent period.

It has been my aim to show, not that much has been done, but that there is a beginning; not that we have entered far into the promised land, the land lying open to the human view so temptingly since the first man looked up to the sky, but that a few pathways are being mapped out, along which we may direct a hopeful attack. Our problems now take a more definite form, and even if we are never to solve them completely, let us remember the words of the poet:

" If God held in His right hand all truth, and in His left nothing but the ever ardent desire for truth, even with the condition that I should err forever, and bade me choose, I would bow down to His left, saying, ' Oh, Father, give, for truth is but for Thee alone.' "

(Applause.)

THE PRESIDENT: I am sure that you will all approve of my action if I express the thanks of the Academy to Professor Kapteyn for taking the trouble to come over here to talk to us on this most inspiring subject. I can not imagine anything more inspiring than the structure of the universe. I will give notice, in accordance with the notice already given, that this is the only lecture or address at this session.

(Thereupon at 11 : 35 o'clock a. m. the session was adjourned.)

AFTERNOON SESSION

RECEPTION AT THE WHITE HOUSE AND PRESENTATION OF
MEDALS BY THE PRESIDENT OF THE UNITED STATES

EAST ROOM, THE WHITE HOUSE, WASHINGTON, D. C.
Wednesday, April 23, 1913, 3:30 o'clock p. m.

The company assembled in the East Room of the White
House.

DR. IRA REMSEN: Mr. President, I have the honor to present
to you Dr. Woodward, the President of the Carnegie Institution
of Washington, and a member of the Council of the National
Academy of Sciences, who will in turn present to you those
who are to receive medals on the recommendation of the
Academy.

We thank you, sir, for helping us in this ordeal.

DR. WOODWARD: *Mr. President, Ladies and Gentlemen:*
In compliance with the wishes of Professor James Craig Watson,
a member of the National Academy of Sciences, and an eminent
expositor of celestial mechanics and orbital astronomy and in
accordance with the provisions of a fund bequeathed by him to
the Academy, a gold medal and an accompanying honorarium
are awarded from time to time for meritorious work in astronomical science. Such testimonials of appreciation and approval
have been hitherto bestowed upon Benjamin Apthorp Gould,
Edward Schönfeld, Seth Carlo Chandler and Sir David Gill.

It is now the decision and the pleasure of the Academy to make
an award to Professor Jacobus Cornelius Kapteyn of the University of Groningen, in recognition of his bold and penetrating
researches in the problem of the structure of the stellar universe.
Out of the apparent chaos of the stars he has shown us those elements of order and system which are fundamental to successful
procedure in the solution of this grand problem. To an indis-

pensable keenness of perception and capacity for generalization he has added the equally essential qualities of courage, optimism and imagination, all disciplined by the rigors of patient observation and refined analysis. Pre-eminent as one of the pioneers in this domain of astronomy, his work is no less praiseworthy for its intrinsic merit than for the fruitful stimulus it has given his colleagues everywhere in the same domain. He has thus helped to make a science always peculiarly cosmopolitan still more beneficent than it has hitherto been in promoting that fraternity of interest which leads to international amity and comity.

It is therefore the pleasant duty of the Academy, on this occasion of its jubilee anniversary, and in the presence of this assembly, to request the President of the United States to convey the Watson Medal and honorarium to Professor Kapteyn.

PRESIDENT WILSON: Doctor Woodward, I feel very much at home in a company made up like the companies in which I have been accustomed to keep a straight face. Notwithstanding my knowledge of the circumstances and although I have long lived in an atmosphere of studies of this sort, I have discovered that the atmosphere is a non-conductive medium and none of the learning of this sort has reached me. I can therefore speak with perfect disinterestedness with regard to matters of this sort.

But, speaking seriously, sir, it is a peculiar pleasure to me to act in the name of this great society so long associated in an advisory capacity with the Government of the United States, in bestowing this medal upon so distinguished a recipient. Professor Kapteyn, I take great pleasure in presenting it to you.

DR. WOODWARD: For the promotion of astrophysical research and discovery and in memory of Dr. Henry Draper, a member of the National Academy of Sciences, himself a pioneer in the newer science of astrophysics, his widow, Mary Anna Palmer Draper, deeded to the Academy, in 1883, a fund the income from which is used for the award of a gold medal in recognition of work already done, or for the direct promotion of investigation in this science. Draper Medals have been awarded hereto-

7

fore to Samuel Pierpont Langley, Edward Charles Pickering, Henry Augustus Rowland, Herman Carl Vogel, James Edward Keeler, Sir William Huggins, George Ellery Hale, William Wallace Campbell and Charles Greeley Abbot.

On this occasion the Academy announces its award of the Draper Medal to M. Henri Deslandres, Director of the Astrophysical Observatory at Meudon, France.

More than a century ago the annals of science were rendered forever memorable by the contributions of three illustrious Frenchmen; Lagrange, Laplace, and Lavoisier. They set standards of excellence for investigation and for exposition which have since served as models for the world of science. It is with the spirit and with the thoroughness of these masters that M. Deslandres has pursued the rapidly expanding fields of astrophysics during the past quarter of a century, until he has become one of the pre-eminent leaders of this science. His work of necessity extends to a wide range of activities; embracing physical observations of solar eclipses; investigations of electric and magnetic phenomena of the sun; contributions to our theories of auroras, new stars and nebulæ; studies of spectra of the sun, stars and comets, and photographic determinations of the rotation periods of planets. His researches with and improvements of the spectroheliograph are especially noteworthy, for he has thus been able to show the structure and the physical nature of the sun's gaseous mass at varying depths from its turbulent surface. In recognition of the high merits of these contributions to science and in commendation of his effective organization of the Meudon Observatory the Academy bestows upon him the Draper Medal.

In the absence of M. Deslandres, the President of the United States is requested to deliver this medal to His Excellency the Ambassador from France for transmission to our eminent colleague.

PRESIDENT WILSON: Your Excellency, I am very sorry that the distinguished recipient of this medal could not be present, but it gives me a great deal of pleasure, as the representative of the United States, to place it in your hands, sir, the distinguished representative of a nation which has always been our very much esteemed friend. (Applause.)

DR. WOODWARD: A year after the death, in 1910, of Professor Alexander Agassiz, then President of the Academy, his friend and distinguished colleague, Sir John Murray, conveyed to the Academy a trust fund for the purpose of founding a commemorative medal to be known as the Alexander Agassiz Gold Medal and to be awarded from time to time to an individual of any country for meritorious contributions to the science of oceanography. On this occasion the Academy announces its first award of this medal to Dr. Johan Hjort, Director of the Norwegian Fisheries.

Long eminent for their contributions to knowledge in all branches of science, it is a noteworthy fact that in recent decades, our Scandinavian colleagues have proved themselves pre-eminent in the sciences of the atmosphere and hydrosphere of our planet. They have been especially happy in applying recondite theory to effective account in meteorology and in hydrography, proving in the most conclusive manner that there is no antithesis between sound theory and safe practice. Conspicuous amongst those who have contributed to this advance is the Director of the Norwegian Fisheries, who combines in a singular degree the capacities of the investigator, the administrator, and the luminous expositor. His encyclopedic work, in collaboration with Sir John Murray, on "The Depths of the Ocean," based largely on the researches conducted by him on the recent voyage of the Norwegian steamer *Michael Sars* in the North Atlantic, is at once a compendium for the expert, a source of instruction for the student and a foundation for future progress in this domain of science. In appreciation of these achievements, so important in abstract research and so praiseworthy for their economic applications, the Academy makes to him this first award of the Alexander Agassiz Gold Medal. In the absence of Dr. Hjort, the President of the United States is requested to convey this medal to His Excellency the Minister from Norway. (Applause.)

PRESIDENT WILSON: Your Excellency, it would have been most agreeable to us if we could have seen your fellow countryman who has looked into matters so deep and unstable, but it is a great pleasure in his absence, to put this medal into your hands for transmission to him, and to express to you our appreciation of his work and of your own relations with the United States. (Applause.)

DR. WOODWARD: Our late colleague in the Academy, General Cyrus Buel Comstock, member of the Corps of Engineers of the United States Army, won distinction as Chief Engineer on the Staff of General Grant during the great Civil Conflict. But in the pursuit of his arduous vocation he found time also for the cultivation of science, and he is not less distinguished for his contributions to geodesy than for his service in the evolution of our Commonwealth.

His devotion to physical science is witnessed in his last will and testament, by which he conveyed to the Academy a fund the income from which may be used for the promotion of researches in electricity, magnetism and radiant energy. Under the terms and conditions of this fund the Academy now makes its first award under the designation " Comstock Prize " of the sum of $1500 to Professor Robert Andrews Millikan of the University of Chicago.

It is a far cry from the adumbrations of Democritus and Lucretius to the modern doctrine of atomicity. But the demonstration of this doctrine, dimly foreseen more than twenty centuries ago is the greatest achievement in physical science of the past two decades, and one of the greatest in the history of science. It is now proved not only that what we call gross matter is atomic, but that what we call electricity has also a granular or atomic structure. With rare acumen and with rare experimental skill Professor Millikan has furnished the most direct and the most convincing proof of the existence of electric atoms or elements.

He has shown how to count these elements in any small electrical charge; he has rendered them almost tangible by showing in the clearest manner their visible effects, he has determined with superior precision the fundamental constant represented in the electrical charge of these atoms; he has demonstrated the equality in electrical charge of the positive and the negative ions in ionized gases, and he has made important additions to our knowledge of the molecular constitution and the kinetic phenomena of gases.

For these contributions to knowledge and for the original and refined methods of research he has developed and so successfully applied, the Academy honors him with this first recognition of superior merit as provided by the founder of the Comstock fund.

PRESIDENT WILSON: Professor Millikan, in presenting this, may I not express my gratification that you have brought us out of the adumbrations of Lucretius into modern life, and say that I envy you the practice of a science so incalculable and exact? (Applause.)

Whereupon at the conclusion of the above ceremony, the President escorted his guests into an adjoining room to partake of refreshments.

THIRD DAY

BALL ROOM OF THE NEW WILLARD HOTEL, WASHINGTON, D. C.

Thursday Evening, April 24, 1913.

At the conclusion of dinner, the assembled company was called to order by President Remsen.

PRESIDENT REMSEN: Gentlemen, I have received information to the effect that the members of the Academy and guests are tired of my voice (cries of " No, No "), having listened to it for three days, and under these circumstances I am going to ask my friend and colleague, Dr. Woodward, to act as toast-master on this occasion. (Applause.)

DR. WOODWARD: *Mr. President, Mr. Vice-President of the United States, Mr. Ambassador, Honored Guests, Members of the Academy:* It has been well said that wheresoever and whenever three or more Americans are gathered together, there and then you will find also a president, a secretary and a treasurer (laughter), and soon after that, talking begins.

It is a singular fact, however, that the National Academy of Sciences, on such occasions as this, has distinguished itself by a minimum of talk.

Our President, and the members of the Academy who are old enough, will remember that it has been our uniform custom to have but a single toast, and that drunk to the Academy. On this occasion, however, which happens only once in fifty years, I think we may depart from our usual custom, and I am going to ask you to rise and drink, with me, to the health of the President of the United States. (Applause.)

(The assembled company stood and drank the toast proposed.)

DR. WOODWARD: It is the function of a toastmaster to be at once as polite as diplomatic usages will permit, and to be a dictator.

It was not my expectation that I would be thrown into this position tonight; but it happened that it fell to my lot to engage the speakers for this occasion, and I will take the liberty now to remind them that they were cautioned that their speeches were to be rather narrowly limited, to something between five and ten minutes. In compensation for this restriction, it should be said, however, that although we have some set toasts—it is easy to make those, you know, if you have a typewriter and an efficient secretary—we shall not confine any man to any particular subject, but we do impose the restriction of a time limit.

Some of the distinguished gentlemen with whom I have been in correspondence have protested that it would hardly be fair to them (not being professional men of science) to be called upon to speak concerning scientific subjects. It should be said with respect to this that they probably possess peculiar qualifications for such a task. You will remember that our late distinguished statesman and jurist, William M. Evarts, on an occasion which was of great distinction in the history of American science, namely, the occasion of the great Tyndall dinner held in New York some years before the younger gentlemen here were born, was chosen as presiding officer; and he said that probably the only reason why he was chosen was that people recognized his impartial ignorance of all branches of science. (Laughter.)

I make these prefatory remarks before calling on the Vice-President of the United States, whom we are delighted to have with us this evening, and prior to announcing that he will speak to us, if he wishes, on the relations of science to government.

I take great pleasure in introducing to you the Vice-President of the United States.

(The assembled company rose in acknowledgment of the introduction, amidst applause.)

SPEECH OF HONORABLE THOMAS R. MARSHALL

Vice-President of the United States

THE VICE-PRESIDENT: *Mr. Toastmaster and Fellow Scientists* (laughter) : I thank the Toastmaster for his frankness in reading my scientific biography; and yet I think that, perhaps, I am as much a scientist as some men who pretend to know more about science than I do, for I may declare with (I believe it was) John Stuart Mill, that much of the discussions of life arise over the fact that men do not understand each other's language.

I think that nothing in the world could be finer than to belong to a body of scientists, and to pretend to be " scientific," because it is such a variable word—it is like our currency, so elastic; it embraces not only what you know, but what you think you know, and what you ought to know. (Laughter.)

And so, almost anybody in America can claim to be a scientist.

There is, although you may doubt it, gentlemen, a science of government in America that consists in putting the other fellow out and getting in yourself. (Laughter.)

Among some of the civilized peoples of the world, however, there are certain scientific rules that have much to do with government; but, seriously, I am glad to be present at the Fiftieth Anniversary of the National Academy of Sciences, and I am glad to say that, much as it has accomplished in the past, I believe it can accomplish much more in the future for the Republic.

Perhaps Mr. Edwin Chadwick (I think the distinguished Ambassador from England will admit) was a great Englishman, and Mr. Chadwick—while I do not quote him accurately—has said that there was hardly any subject about which legislators were willing to legislate where, when investigation was made, it was not found that they were willing to legislate on the wrong side of the question. (Laughter.)

I think he produced some illustrations of the falsity of the views which are held in political life when the trained men have found the facts. I think he disclosed that the old Malthusian doctrine of the glut of population—that God sent wars and pestilence to get rid of the glut of population—proved to be not true in view of the fact that after every war and after every pestilence, about two children were born to every one that died in the pestilence or had been killed in the war.

That poverty was the result of glut of population was found not to be true in England when the English people gained an opportunity to go to work or go over to the colonies. So they went to work, rather than come to America.

There is only one objection that we can have to scientific investigation, and scientific investigation which goes to the betterment of government, whether in legislative or judicial matters; and I may be permitted to voice that objection.

I think it is not applicable to this National Academy of Sciences, but I know that it is applicable to the life of the nation at large, and that objection is that the scientific man very frequently, instead of attempting to serve the state, either in the halls of legislation or before the court, suffers and permits himself to become a partisan and to advocate, not his own clear, definite and fixed opinion upon the subject, but an opinion which is colored by his having been retained by one side of the case or the other. (Applause.)

I know of no people who have reached a right conclusion with reference to the value of science in the courts except the people of Norway and Sweden. There the expert is the officer of the government, and after the facts have been found by the court, the expert is called in, and he takes his scientific knowledge and information and settles the case after the facts have been found.

In the administration of justice in America, none of us country lawyers could ever afford to hire as good men as you are, because, perhaps, you would not be hired, and because, perhaps, we could not pay the price if you would be hired; but in a smaller degree, your little duplicates, your small children, who are in all the villages and cities and towns of America, from fifty dollars to five hundred dollars, will get a man as expert to testify on any side of any subject in any court in America. (Laughter.)

Just now, gentlemen, I would not want to say that this Government needed any advice from scientific or other men outside of the Democratic party (laughter), unless it might be how to obey the civil service law of the land and satisfy the desire of the average Democrat for office. (Laughter and applause.)

If, either by telescope or microscope or any other " scope," science can determine that question for my brother upon whom rests the burden of government, there will be no reward which Congress will not be willing to bestow upon the eminent gentleman who finds it out.

I say that, just now, it may not be necessary for the United States of America to call for advice; and yet, a plain, untutored, uncultured man—myself—I have both a veneration for you and a fear of you; a veneration for your devotion to subjects which seem to be abstruse, which have but little to do with statecraft, so far as the world sees it, and but little to do with the accumulation of wealth and property, and a fear of you that, in your investigations, you may discover that the economic policy of the Democratic party is wrong. (Laughter.)

It is good for a man, however much he falls short himself in the splendid ideals of life, to look into the faces of men who have lifted themselves above the petty cares, the turmoil and the vexations of the ordinary business and professional life, who have dedicated and consecrated themselves to the finding out of what is truth, who want, twenty centuries after carping Pilate asked the question, to make an answer to a still further inquiring world.

It is good to look upon a body of men who do not stand as men stood in the early ages of the world, looking at and believing every thing they saw, but rather as men upon whose shoulders there now rests the burden of the centuries, until their backs are bowed and each stands forth as an interrogation point, asking, " Why, why, why " of everything that is seen.

It is a splendid thing to face a body of men such as you are and to bid you God speed, peace, contentment, long life, and to pray you to still continue to have that sacred devotion to the truth which will freely cast aside everything that you now believe, if you find it to be untrue, and which will joyously grapple the truth, however much it may be inimical to your own personal desires.

Gentlemen, I thank you. (Loud and continued applause.)

DR. WOODWARD: Gentlemen, although as the Vice-President has well said there has been recently a marked change in the politics of the United States, there are some of us who can remember that, only a few years ago, there was a very popular administration led by our friend Theodore Roosevelt. (Applause.) During his administration I heard one of our fellow citizens remark, " Next to Roosevelt, there is one man who is the most popular man in the United States." A few years later we had a somewhat less popular administration, under President Taft; and the same person repeated the same remark. He said, " Next to President Taft, who has been very popular, there is one man, not a citizen of the United States, who is the most popular man in America."

Now, quite recently, since the incoming of the present administration, the same remark has been repeated.

I hardly need tell you to whom I refer. He is the man who has taught Americans to understand themselves better than any other man. He is the man who has won our affectionate regard, and the man whose return to his native land—I will not say we deplore—but whose return we hope will not be for long. I refer, of course, to the British Ambassador, the dean of the diplomatic corps, and Fellow of the Royal Society of London, and I would suggest, with the largest liberties as indicated before, that he may perhaps be kind enough to speak to us of foreign academies and societies. (Applause.)

SPEECH OF THE RIGHT HONORABLE
JAMES BRYCE
Ambassador from Great Britain

AMBASSADOR BRYCE: *Dr. Woodward, President Remsen and Gentlemen:* I am very much touched by the kind words in which my old friend, Dr. Woodward, has introduced me to you, and I am more than grateful to you for the way in which you are kind enough to receive me. It does make one happy to be so received and to be assured that one has not lived in this country six years without having acquired some friendliness on the part of its people.

But, apart from that, gentlemen, I stand before you this evening as a rather unhappy man, because it is the last evening on which I am likely to have the privilege—at any rate, in an official capacity—of meeting an audience of American men of science.

One of the most delightful parts of my sojourn in Washington has been my intercourse with your men of science. There is not any city in America—I doubt if there be any city in Europe—where so many men of eminence in science are assembled as live in Washington, and the gatherings which you have here, when the men of science from the whole of your wide country come together, have been among the most delightful experiences that I or any Briton has had in this country. I have had it also in Philadelphia and in New York, and I have had the pleasure of making the acquaintance of your men of science in many journeys all over the country; but this, after all, is the focus to which is gathered most of the scientific lights and leaders of the United States at stated intervals when you come together here.

And I can assure you, gentlemen, there is nothing I shall look back to with more pleasure, in so much of life as remains to me, as to the friendships I had formed with your scientific men and the inspiration I have derived from the ardor and energy with which they pursue the studies to which we are all so much debtors.

Dr. Woodward has suggested that I should say something about foreign academies, but my knowledge about foreign academies is, really, practically confined to my own country, for, whenever I have traveled abroad, it has rather been among the historians than among the men of science that my work has lain.

However, I should in any case feel a little doubtful about venturing to talk about scientific academies, knowing that, whatever else " science " means, Mr. Vice-President, it is supposed to mean knowledge; and if a man feels that he does not know a thing, scientific people are the last to whom he should address his remarks.

I received at Oxford my literary education, and I remember " education " being defined by a very eminent professor there, who said:

" What our Oxford education does is to teach our men to write plausibly about subjects they do not understand " . . . .

an art which we were in the habit of exemplifying by immediately beginning to write for the journals and reviewing books —whose authors knew infinitely more about their subjects than we did—in a very superior manner, an experience which, however, is not confined to England.

The Vice-President said, gentlemen, that he regarded men of science with fear and veneration. I share those feelings. I have veneration for the lofty and disinterested spirit which you bring to your work. I have fear for the enormous power you exercise.

You are really the rulers of the world. It is in your hands that lies control of the forces of activity; it is you who are going to make the history of the future, because all commerce and all industry is today, far more than ever, the child and product of science; and it is you who make these discoveries upon which, when they are applied by industry, the wealth and prosperity of the world depend. It is in your hands that the future lies, far more than in those of military men or politicians.

But I have another feeling besides fear and veneration. It is that of envy. I envy you your happy lives. Compare your lives with the lives of any other class. If the Vice-President will permit me, I think the life of a man of science is a great deal happier than the life of a politician or the life of a statesman, who, as we know, is many pegs above the politician (laughter), because the politician is occupied, as the Vice-President has said, in endeavoring to promote the interests of his party and not the interests of his country; and I discovered, during my experience in the House of Commons in England, that a legislative assembly is the worst place in the world for the discovery of abstract truth. (Laughter.)

Or, take the case of the lawyer. So far from seeking to discover the truth, in one-half of the cases which he conducts, he is endeavoring to obscure the truth. (Laughter.) Or, even take the case of the artist or the literary man, who has a subject to work upon, delightful and interesting in itself, in evoking from the stone, or by colors, shapes or forms of beauty, which will far outlive him. But these forms of beauty will profit him very little if they do not commend themselves to the popular tastes, and he is constantly under the temptation of doing something less good than he wishes, in order to meet the tastes of his patrons.

It is the man of science who has the really happy life. He is engaged in the discovery of the truth, and nothing but truth. The applause of the multitude is nothing to him. He is working for a mistress more exalted than any popular assembly or any multitude that we can conceive of. He is working for Truth herself, and for the future. He is consecrating his efforts to the highest task that God can lay before man, and in that he needs nothing but the sense of what he is adding to the sum of human knowledge, and he has the incomparable pleasure of feeling that the more he knows, the more the immense ocean of knowledge stretches itself out before him. The further he outlines any path into the untrodden solitudes of ignorance, and the more he blazes those paths and makes them paths of knowledge, the more he sees other paths branch out before him, leading further and further away into the realms which others after him will traverse.

In these things, friends, there are elements of pleasure and delight, elements also of independence, which I think no other profession can equal.

I was tempted to add one other charm which your life has. It is the charm of poverty. (Laughter.)

I have sometimes felt inclined to wish, Mr. Vice-President, that Congress was a little more liberal to the scientific men who are working for Uncle Sam. (Applause.) But perhaps they are to be congratulated on being free from those temptations which beset wealth. Poverty, like other things, is good if you have not too much of it (laughter), because it saves one from the temptation of forgetting the end for the means, the temptation to which most of us, and, above all, those who are in search of wealth, succumb. You keep the end always before you, and you proportion your life to that end.

Still, I think you might, with advantage, not only to you, but, what is far more important, to the whole country—and it ought to be possible in a wealthy country like this—provide upon a more ample scale for those who follow science, and give science a more exalted position, by freeing the scientific man from any thought of domestic anxiety.

You enjoy in this country—I speak here of particular branches of science—some things which we, in England, greatly envy. Think of what the geologist or the botanist has before him here! We have been working for one hundred and fifty years upon the geology, and for more than that upon the botany, of our little island; but here you have the whole continent open to you, and any man of science on these subjects can make a reputation for himself by new work in new fields, such as is impossible for us in outgrown Europe.

Gentlemen, one word I venture to say about the scientific bodies of the continent of Europe. We have, in the Royal Society, the oldest of those bodies, and one which, I think, has always maintained the level which it took in the great days when Isaac Newton was one of its members; and now there has sprung up all over Europe a host of other bodies pursuing science and following it into those infinite ramifications which modern science has discovered. Everywhere there you are welcome. One of the most delightful things of science is that it knows no divisions or allegiance to nationality. It is a republic in which there is no passport to greatness, except service and genius, and it is a republic of which everyone is a citizen, and where everyone has equal rights in every part of the world. (Applause.)

That has always been our tradition in England and in our Royal Society; and I know it is your tradition here, and I know what hearty welcomes you have always given to our men of science when they have come over here, and how refreshed and invigorated in spirit they have been when they have gone back to their own country.

Gentlemen, I can wish nothing better for any of us than that these comings and goings will be frequent, and I can assure you that it will always be a pleasure to the scientific men of England and Scotland to welcome you to their societies and to all their gatherings and universities. I hope that, more and more, these meetings will take place, and I can assure you that all you have achieved and all that you are achieving in so many ways on so many different lincs for the advancement of knowledge, for the extension of human power that comes through knowledge, is followed with gratitude and admiration by the scientific men of Great Britain. (Applause.)

VICE-PRESIDENT MARSHALL: Mr. Ambassador, your eminent services to the Republic are such that I feel that I dare address you, and I want to say that if these scientists will fix the status of a Vice-President and will convince the general government that he has a right to live, whenever they do that, then, Mr. Ambassador, I will try and do something for the scientist. (Laughter and applause.)

DR. WOODWARD: The fields of science are many, and they are coming to have such numerous ramifications that the individual can no longer, in general, hope to be acquainted with more than his own specialty.

But we have with us tonight one who marks a very noteworthy exception to that rule, a man who, long before some of us were born, had acquired distinction in science.

I refer, as you well know, by this time, to our eminent colleague, Dr. Silas Weir Mitchell, the oldest living member of the Academy, but who is still so young that he is the most suggestive member of the Academy in regard to new things and new projects for observation and experiment. But, in addition to these numerous fields with which he is acquainted, and in which he is an expert, he has long delighted us with his graceful fancies and his historical instruction, produced in his works of fiction, so-called.

I am going to call upon Dr. Mitchell to tell us something of the reminiscences of the Academy, which he can do better than any other member of the Academy. (Applause.)

8

SPEECH OF DR. SILAS WEIR MITCHELL

DR. MITCHELL: *Mr. President, Mr. Vice-President, my Brothers of the Academy:* I am, I presume, the victim of the after-dinner hour, as usual, and am well aware of the treachery of the tongue, and much prefer the loyalty of the pen. I have, therefore, deliberately put on paper that which I want to say to you tonight, feeling that it will be much more probable that I shall interest you than if I trusted to my unassisted words.

I am, I presume, indebted to the liberal forbearance of time for the honor of being asked to speak to you this evening. It does not find me in the careless mood of after-dinner gaiety, nor is it possible to escape altogether from personal remembrances, which elsewhere than at this friendly board might entitle me to be relegated to what Disraeli called the " fatal time of anecdotage."

My diploma is dated August 25, 1865, three years after our foundation. It is signed by Dallas Bache, Wolcott Gibbs and Louis Agassiz. Since then, one hundred and thirty-six of our fellowship have come and died, with an average duration of academic life of more than eighteen and a half years—very many with far less. This makes clear that in those earlier years our additions were of men older than those we elect now.

At present the liberal endowment of research opens the way to distinction for younger men, unembarrassed by the time-killing need to preach science as well as to practice it.

Between the mere words of our record—*elected—deceased*—you, who are familiar with our history, may read much that is written clear on the roll of scientific achievement.

Here are they to whom, from the depths of space, were whispered in the night watches its long hidden secrets. There, too, are those who, in the silence of the laboratory, rejoiced in the fertile marriage of the elements, or they who, like confessors, heard from dead bones or rock or flower the immeasurable history of the silent ages of earth.

One might linger long over many of these lives whose interests were so remote from thought of the commercial gains their unselfish work made possible. But there are other compensations, and there are men here today who are aware that there is no earthly pleasure more supreme than to find disclosed some secret of nature unknown before, save to Him who set in motion the complex mechanism of the universe.

The later life of the merchant and the lawyer loses vitality of normal interest as age comes on; not so the man of science. The eternal love of nature is his—a mistress of unfading charm.

I remember once that, at my table, someone asked that ever happy naturalist, Joseph Leidy, if he were never tired of life. "Tired!", he said, "Not so long as there is an undescribed intestinal worm, or the riddle of a fossil bone, or a rhizopod new to me." (Laughter.)

My first remembrance of an Academy meeting is of 1866. We met in a Smithsonian room. There were not more than fifteen present. Professor Henry was in the Chair.

I remember Benjamin Peirce, Wolcott Gibbs and Gould. Agassiz sat on one side of me, and on the other Coffin. It was all very informal. The first scientific paper was by Professor Peirce, who for twenty minutes occupied us with algebraic formulas and mathematical figures, until he turned and said that he had got out of the region of material illustration, and so led us on through the endless equations in which I had lost myself at the very outset.

Agassiz turned to me at the close and said, "Were you able to follow him?" I said, "No; I can not do a sum in the Rule of Three without trying it over two or three times." Upon which the delighted naturalist added, "Ni moi non plus." Professor Coffin remarked, "He was traveling with Seven-league Boots over a country across which I should have to crawl."

Some of this was quite audible to Peirce, who said that the only thing required was more careful attention than men were willing to give to the great science of mathematics, and that our incapacity to understand and follow him was due to our want of proper education.

He was succeeded by Agassiz, who made the first announcement of his discovery of the additional heart found in the tail of the young of the salmon.

I recall very little else about these delightful people, except that they—all of them—were not only in the peerage of science, but also companions as socially interesting as they were learned.

Perhaps the most pleasant remembrance I have of all is of Louis Agassiz and Joseph Henry. The former was good enough to take a great interest in some of the animal physiology with which I occupied the rare leisure of a hardworked young doctor. His enthusiasms were shown in odd ways at times.

On one occasion he was staying with Professor Frazier, and dismissed me on the front step one slippery day in February. I had got some distance from him when he came after me in haste, sliding over the pavement. " I did want to say to you one thing. Are you acquainted with the opossum? " I said, rather confused, " No." He said, " I advise you to acquire a physiological friendship with the opossum. He is a mine of physiological wealth." (Laughter.)

Jeffries Wyman, who was elected in 1872 and died in 1874, was another who held a place in my most honoring regard. He resembled Joseph Leidy in that splendid magnanimity and unselfishness which contrasted so agreeably with the disgusting quarrels, happily rare, which sometimes arose among men of science.

As you have made me speak here, I am forced to say something of myself, and hence this anecdote of Wyman. I had written him word of the discovery I had made of the chiasm of the superior laryngeal nerves in the chelonia—that is to say, turtles—and it greatly excited him, especially my prediction that it would be found in serpents and probably in birds. A year afterwards he sent me a large bundle of illustrations and descriptions of what he had found in other classes than the turtle, and insisted that I should use them in the second paper I was about to print, stating that they would not have been discovered had it not been for my predictive aid. Of course, I declined this help; but it was characteristic of the noblest form of the scientific mind.

You will, I trust, pardon me if I close this long talk with a few too personal words about the much loved first director of the Smithsonian Institution, first of the men who sacrificed to that Institution a scientific career. When a boy about fifteen years old, I was sent by my father to Professor Henry at Princeton with some glass apparatus, which could not otherwise be sent without danger of breakage.

He met me at the station, took me to the house, and spent a part of the next morning endeavoring to explain to my bewildered youth the experiments he was making in the transmission of electric signals. I was overcome by the unwonted attention paid to a boy of my age, and expressed myself so warmly that he laughed as he bade me good bye, saying: " Well, life sometimes gives one a chance to return little favors, and perhaps some day you will have an opportunity to oblige me."

Long years passed by, and sometime in the beginning of 1878 Professor Henry asked me to come to Washington and advise him. After a thorough examination of his case, he asked me plainly if he was mortally ill. I said, " Yes." Then he asked how long he had to live, but I could not set a date. He said, " Six months? " Hardly, I thought. He died in May of that year.

As I arose to go away, his carriage waiting, he said: " I have yet to discharge my material obligation. How much am I in debt to you? " I replied, " You are not in debt. There are no debts for the Dean of American Science."

He was much overcome, and said: " I have always found the world full of kindness to me, and now here it is again." I could only say: " You do not remember, sir, that once you said to me, a boy, when you had been very kindly attentive to me and I had tried to express my obligation, that perhaps a time might come when I could oblige you. If this obliges you, my time has come." And so we parted.

I may add what some of you already know, that Alexander Agassiz wrote me he had intended to return home early from Europe, in order to give a dinner such as we are having here tonight. He died on the way over, and his letter reached me after his death—strangely enough, the fourth letter I have received from men who were not alive at the time their words reached me.

My talk has been of men dead long ago, but I should be ungrateful to the longest friendship of my life if I did not pause to remind you of our latest loss in John Shaw Billings. He was a man of too many competencies to allow of even allusive comment here.

Few men have been better loved or had so enviable a capacity to convert mere acquaintance into friendship.

It is difficult, sir (addressing the President), for a man as old as I am to talk in the gay after-dinner mood, and if I have been too sombre and too personal, I trust that I may not have been guilty of the social crime of having been uninteresting. (Prolonged applause.)

(At this point, 11:10 o'clock p. m., Ambassador Bryce took leave of the assembled company, which rose in acknowledgment of his departure.)

DR. WOODWARD: On an occasion like the present, we are naturally disposed to magnify the good works of the Academy. But attention should be recalled to the fact that the Academy is not the only scientific society in America, and by no means the oldest. It is an interesting phenomenon, that in recent decades, America has given birth to many scientific societies. But, behind them all, is the oldest of our American societies, namely, the American Philosophical Society Held in Philadelphia for the Promotion of Useful Knowledge; and we have the pleasure this evening of having with us the distinguished and honored President of that Society, Dr. Keen, who will tell us something of the American Philosophical Society, our oldest sister society in America. Dr. Keen. (Applause.)

SPEECH OF DOCTOR W. W. KEEN

DR. KEEN: *Mr. President, Mr. Toastmaster, Mr. Vice-President, and Gentlemen:* It is a great pleasure to me to bear the greetings of the oldest scientific society in America to the National Academy of Sciences, and I am sure I voice their sentiments when I say that we wish you—and we could express no better wish—an equal degree of prosperity and success for the future to that you have had in the past.

When I received Dr. Woodward's kind letter, if he had asked me to get busy in a professional way and disembarrass the members of the National Academy of Sciences from the peccant appendix which troubles you all, and which now is working very badly within the dim recesses of your interior, you can imagine how gratefully I should have accepted the task; but when I discovered that he had asked me to make an after-dinner speech I was reduced very much to the condition of the Irish soldier who applied for a pension:

" Where were you hit? " said the surgeon.

" The bullet went right through here," he said. " No, my man, that would be impossible, for it would have gone straight through your heart, and you would have been a dead man immediately." " Sir," he said, " when I was hit I was so scared that my heart was in my mouth." (Applause.)

If, however, my dull tongue commits as serious wounds upon you as my sharp knife might have done, I am sure that you have brought it upon yourselves.

The American Philosophical Society had its origin 186 years ago. In 1727 when Benjamin Franklin, who, my friend at my right has wittily said, was not born in Boston, but was born in Philadelphia at 17 years of age (applause)—when he was 21 he gathered around him—an extraordinary thing for so young a man—a number of congenial spirits who were interested in scientific matters and formed his famous Junto. Sixteen years later, in 1743, from this society, informal as it was, arose in formal manner the American Philosophical Society. It is an extraordinary thing that so young a society as this should have adopted the Emersonian method and hitched its wagon to a star, by taking as its model the Royal Society, the oldest and greatest of all the scientific. societies of the world, and it has steadily pursued this method and conducted its meetings and its purposes after the same fashion. In one respect, however, it has departed from the course of the Royal Society, in that it has retained not only the biological and the physical sciences, but also the humanities. Accordingly, we have with us always a number of men who are distinguished in other branches than science, properly so-called. Any of you that can be present on our Thursday afternoons or the annual joint meetings will, I am sure, be delighted and pleased with the fine papers that are read during that session of the society.

Among the early members that Franklin gathered around him he mentions by name a physician, a botanist, a mathematician, a mechanician and, under one extraordinarily vague but comprehensive title, one "general natural philosopher." While the others are mentioned by name, the name of this extraordinary genius is buried, unfortunately, in oblivion. We have always had very close relations, naturally, with the University of Pennsylvania, and from it has grown very largely our membership. Moreover, the very first Secretary of the Philosophical Society was the energetic Provost of the University of Pennsylvania, William Smith, who, on one occasion, for an alleged libel upon Quakers was thrown into jail, and, it is absolutely asserted, taught students through the bars of the jail. Moreover, my immediate predecessor in the presidential chair was a distinguished chemist who is now the Provost of the University and a member of this society, Professor Edgar F. Smith.

In 1769, occurred an event of the greatest scientific importance, the transit of Venus. David Rittenhouse, who was the astronomer of the society, imported some instruments from England and also for the observation constructed a clock which, after 144 years, still stands upon our walls and marks with precision, from which no appeal can be taken, the twenty minutes allowed to each paper.

Rittenhouse's observation of this transit were the only valuable ones taken at this time, for, unfortunately for the European observers, they had cloudy weather. Rittenhouse had beautifully clear weather; and for 105 years, until the next transit occurred, which most of us remember in 1874, his observations were the basis of the astronomers for that period. For this transit Rittenhouse built three observatories, one of them immediately in the rear of the present building of the American Philosophical Society in Independence Square itself; and it is an interesting fact that it was from the balcony of this little observatory that, on the 8th of July, 1776, the first public reading of the Declaration of Independence took place. Moreover, we have in our archives, not only the chair upon which the Declaration of Independence was written by Thomas Jefferson—his library chair—but we have also the first rough draft of it with the corrections or suggestions of law, and we have also one of the rarest of historical documents, a broadside that was posted on the walls and fences of the town in 1776; and all of these precious relics, with many others, are exposed today, in a combustible building, erected in 1887, to the risks of fire.

Of course the Revolutionary War broke up, very largely, our scientific meetings, but they were resumed immediately after the war, and at intervals, in fact, during the war.

In 1790 occurred the death of Franklin, and five years later the death of Rittenhouse. Eulogies were pronounced before the Philosophical Society upon both of these distinguished men, and these meetings were attended by the President of the United States, by both houses of Congress, by the diplomatic corps, by the judges, and the various public citizens who were invited.

Jefferson, I might say, was our president for eighteen years, and incidentally I might remark that during eight years of those eighteen he was also filling the minor office of President of the United States. (Laughter.) It is the unique distinction of the Philosophical Society that we have furnished eight presidents of the United States, from Washington to Wilson. (Applause.) They were elected to the Philosophical Society not because they were presidents, for their election preceded their presidency, but we fondly think, in the Philosophical Society, that they were elected presidents because they were members of the Philosophical Society   (Laughter.)

And, I might remark that the recent election of President Wilson has left all of the other members of the society in a very receptive mood.   (Applause.)

In 1906 we celebrated the 200th anniversary of the birth of Franklin, a very notable celebration, when addresses were made by the most distinguished men of our own and other countries, including the Ambassador who, unfortunately, has just left us. There were also three other things that occurred at that time that are rather notable. We published the calendar of Franklin papers edited by our accomplished Secretary, Dr. Hayes. Congress struck a medal commemorative of the event, of which a gold duplicate was presented to the Republic of France, and also a portrait of Benjamin Franklin, which, by the casualties of war, had become the property of Earl Grey, then Governor General of Canada, was restored by Earl Grey in the most generous manner to the United States and hangs today in the White House.

We possess of the writings of Franklin, eighty large scrap book volumes which contain seventy-eight per cent of all the Franklin known correspondence and documents.

We have also the only copy preserved of the two that were made of the Charter of Privileges of the Colony of Pennsylvania, the one belonging to the State having disappeared; and this copy, that belonged to the family of Penn, has found its home in the Philosophical Society. We have also the instruments that were used by these distinguished engineers at that time to trace the boundary line between the states of Maryland and Pennsylvania—the well known Mason & Dixon Line, which, I am glad to say has been wiped off the political map, though it were by the bloody sponge of war.

We have a number of the earliest explorations also among our archives: Dunbar's exploration of the Red River in 1804-1805; the original field notes of the Lewis & Clark celebrated expedition into the great Northwest, the first that ever penetrated to the Pacific Coast, deposited with the society at the request of Jefferson, who had appointed them on the commission, and who was then President; Michaux's Botanical Journal; Muhlenberg's Botany of Pennsylvania; Priestley's ingenious dissertation on Phlogistin, dated 1873; Franklin's Electrical Machine, and other treasures, which time does not permit me to mention. We have always had a very pleasant and close relation of membership, especially with this National Academy of Sciences. Going over the list of the fifty original members I find that forty-one of them were already members of the American Philosophical Society, and only nine were non-members. The present membership of the society is very largely a duplicate of the membership of the Philosophical Society. Many of you, who are fresh from our general meeting last week, I am sure will agree with me that we have not lost our youthful vigor, though we are approaching the end of our second century, and that we bid fair to outlive not only the second but, I trust, a third and more centuries afterwards.

Our two co-ordinate societies move on, hand in hand and heart to heart, in loyal co-operation, for science knows no boundaries, geographical, political or linguistic. Its devotees are diligent in Arctic cold and tropic heat; they flourish under kaisers and kings and presidents. We gather the precious harvest from all peoples, realms and races. We gather not only our harvest, but we scatter abroad the precious seed for the benefit of entire mankind. (Applause.)

DR. WOODWARD: Lest we should forget, in listening to these delightful and instructive reminiscences of our friends Dr. Mitchell and Dr. Keen, from Philadelphia, I wish to remind the society and those here assembled, that there is a very important function of the Academy, defined by legal enactment. The Academy is in a sense the confidential adviser of the United States with regard to matters scientific. We have with us here a distinguished statesman who, if the scope of our organization were less narrow, could easily become a member of the Academy. I refer to Honorable Theodore E. Burton, and will ask him to speak to us concerning legislation and the National Academy of Sciences.

SPEECH OF HONORABLE THEODORE E. BURTON

United States Senator from the State of Ohio

SENATOR BURTON: *Mr. Toastmaster and Gentlemen of the National Academy of Sciences:* It is gratifying to every after-dinner speaker to know that every subject is capable of a great variety of interpretations. In the first instance, taking advantage of this liberty, I wish to dwell upon a theory, involved but accurate, which emphasizes the importance of your calling. I refer to the dependence of popular government and liberty upon the progress of science. That progress begins with the pure sciences, then the applied sciences, then the varied labors of the inventor, the mechanician, and along with these mankind obtains a development under which he realizes his greater importance, the sovereign becomes less and the individual more.

In the early ages there were the Republics of Athens and Rome, but popular government was limited to the size of a single assembly. Widespread enjoyment of freedom under a great Republic like ours was only possible with the development of means of transportation; with the printing press; with the development of manufactures, and of the work of the merchants, under which those who had belonged to a servile class became an important part in the body politic. Kings and sovereigns recognized that it was necessary to give to them their proper place; that they increased the wealth or power of the country; that they could not depend upon conquest alone, but needed the more peaceful fruits of industry.

Francis Baker lived under two sovereigns who wielded absolute power—Queen Elizabeth and James I; but he blazed the way for the development of modern science and invention, and in the reign which followed there was an uprising against tyranny and absolute power.

Civil government, in its better development and greater equality, in all these things which make a man happier and more loyal to his government, has kept pace with the locomotive, with the improved steamship, with the telegraph, with the greater diffusion of news and information, all of which have been made possible by the developments of modern science. We have witnessed a most remarkable phenomenon here within a few years, the renovation in China, which, for thousands of years had dwelt in inertia, and had not merely conservatism, but the absolute lack of progress. What is the reason? The development of the railway and means of communication in that old country; the beginning of manufacturing, the coming of scientific knowledge, the breaking away from old traditions, and the awakening, so that they are keeping pace with modern scientific progress.

In the past ten or twelve years of this century results have been achieved for the betterment of conditions of the people of the United States comparable with the whole century before, although that was probably called the greatest of the centuries. What is the main reason? The wonderful awakening in science; in knowledge, new inventions, new facilities, better methods, all of which have been made possible by the work of men such as those who are here now before us.

I am asked to say something about the relation of this society of science to the government. If there is any one thing that is needed in all our governments, national, state and municipal, it is a higher standard of expert knowledge. We have been getting along on the theory that almost any man is capable of holding office; that almost any man will do for mayor—possibly for governor; but we have come to understand that we must make a study of government in just the same manner that you study chemistry or any other branch of knowledge. One thing which has stimulated this has been the wonderful success of many men who have had only very limited preparation. The philosophers tell us that every effect must have its cause; that a certain amount of labor and of intelligence brings certain results. But I am compelled to say to those before me that I have known so many men who have been made or marred in finance or in politics by pure, sheer luck, that I can not accept the generalization. (Laughter.)

Nevertheless, the fact of occasional success, without adequate education and preparation, should not lead us to overlook the fact that real, substantial success, whether it benefits the community or the nation, must depend upon thorough preparation.

My friend to my right could tell us of how much we have lost in this nation of ours by an inadequate understanding of the extent of our resources, by an insufficient realization of the great treasures of the land put before us, by the possibilities of greater wealth, greater comfort, and happiness from the better cultivation of the soil.

We have been going ahead and enjoying privileges unexampled in the history of the world, until we are coming to the point where, instead of abundance, we are threatened with scarcity. To meet the situation this government and every government must bring to its aid scientific men like yourselves.

I rejoice to see that great Agricultural Department in a measure a great university. I am glad to see a different disposition on the part of legislators in recent years, which leads them to bring to their aid men of the highest preparation and of the best intelligence and knowledge in every branch of science. Sometimes it may seem to you that men who are lacking in qualifications are placed in positions of responsibility where only those who have had the training you have had should be called. But I am satisfied there is an improvement in this regard; and in this time of indifference to the state, when every man seems to be occupied with his profession or occupation, I call upon you to give your very best thought to the welfare of your country.

We are having a time now of seething discontent; we are having a time when those who prepare themselves best for their duties sometimes think that it is useless. The remedy for all this is the intelligent consideration of problems of government by every one of you and I feel that I can assure you that while there have been defects in the past we have not gone altogether wrong; but we have not taken advantage of the intelligence we possess. There will be an improvement, under which this association and all associations of similar scope and similar ability will be called upon to do their proper share, not merely as citizens, but as aids to the Government, if necessary in official positions. (Applause.)

Mr. President, I know the hour is late. I do not wish to speak longer, but if there is any one thing about which I would like to be an apostle, it is the awakening of the average American citizen from his indifference.

President Harrison said: "The framers of our Constitution and no set of men could frame a government so good and so perfect that the best and the most intelligent citizens could go away and leave it alone."

The country needs your services. It needs them in all the varied activities of politics; it needs the merchant; it needs the doctor; it needs the working-man, though he be working in the ditch. Let us all then bow and make our daily prayer: "God bless and save the State." I am sure that this awakening is coming; and we shall pass beyond a purely material plane up to a higher plane on which this great Government, which is the protecting shield over all of us, will awaken a greater degree of love, a greater degree of the attention, of each individual from the Atlantic to the Pacific.

You are celebrating your fiftieth anniversary. The names of many of the members of this association have been written high on the roll of knowledge. The past of the National Academy of Science is certainly secure; the present is secure. I trust for you all in the years to come there may be an enjoyment of the highest degree of prosperity and happiness; that in your investigations you may make such discoveries as shall gratify your ambitions, and in the years to come, and when another fiftieth anniversary has come, this association, as today, may stand among those in the very forefront of scientific investigation and in the progress of American life. (Applause.)

PRESIDENT REMSEN: Gentlemen, it only remains for me to declare this meeting adjourned.

(Whereupon at 11:40 o'clock p. m., the meeting was adjourned.)

A REPRODUCTION OF THE
REGISTER OF THOSE IN ATTENDANCE
AT THE
SEMI-CENTENNIAL ANNIVERSARY

J. C. Kapteyn
E. E. Barnard
Francis E. Nipher
Wm F. Hillebrand
Arthur L. Day
Paul Brockett
Chas B Davenport
J. U. Nef
Charles D Walcott
Geo C Comstock
Wm Bullock Clark
Ira Remsen
E G Conklin
Oskar Bolza
E B Van Vleck
Wm H. Dall

Arnold Hague
F. W. Clarke.
W. B. Scott
Ernest W. Brown
Edward C Pickering
W. M. Wheeler
C. Hart Merriam
James F. Kemp.
Whitman Cross
Wm Davis
J. McK. Cattell
Charles S. Minot
Jos. P. Iddings
John M. Clarke
Edw L Nichols
Edward W. Morley.
Franz Boas
Arthur Gordon Webster

S. W. Stratton

Charles F. Chandler

Charles S. Hastings

Cleveland Abbe

W. W. Campbell

D T MacDougal

Edm B Wilson

Herbert M. Richards

W. T. Councilman

T A Rice

Geog. F. Becker

R. S. Atkinson

T. Mitchell Prudden

H Emry Fairfield Osborn

S. J. Meltzer

F. R. Moulton

Dayton C. Miller

William H. Welch

Geo. C. Merrill

A. O. Leuschner

W. de C Ravenel

Charles Schuchert

T. C. Mendenhall

Edw. S. Morse

B. B. Boltwood

H. M Smith

Wm. Trelease

Robert S. Woodward

M A Pupin

Austin H. Clark

Elihu Thomson

Wm Patten

R A Millikan

Edward Weston.

(page of signatures)

Simon Flexner

Edw. J. Nolan

G. R. Agassiz

Henry B. White

Charles F. Brush

Theobald Smith

Jno H. Brashear

Theo. D. A. Cockerell.

Edgar Odell Lovett

Harry Fielding Reid

Thomas B. Osborne

Wm A. Noyes

Frank B. Taylor

C. L. Shear

H. N. Morse

Charles R. Van Hise

J. C. White

Wm. Eichelberger

Horace L. Wells

Benjamin Boss

Arthur Mann

A. S. Hitchcock

Henry H. Donaldson

W. G. Cobb

W. T. Porter

F. H. Knowlen

Edward W. Berry

Arthur C. Spencer

Alfred H. Brooks

Fredrick V. Coville

Medical Director, H. G. Beyer, U. S. Navy

B. O. Peirce

Julius Stieglitz

Karl E. Guthe

F. W. True

Mrs. F. W. True

E. B. Rosa

Erwin F. Smith

R. W. Wood

Edmund C. Sanford

Barton Warren Evermann

Frank A. Wolff

Mr. & Mrs. V. K. Chesnut

Mr. & Mrs. F. L. Ransome

Mr. & Mrs. W. A. Orton

H. S. Jennings

Mr. & Mrs. T. W. Stanton

J. L. Jayne

Mr. & Mrs. O. H. Tittmann

Miss Mary J. Rathbun

David White

George W. Stose

H. C. Jones

C. N. McBryde

Chonen E. Berry

Theo. Gill

W. J. Humphreys

T. Wayland Vaughan

W. Lindgren

Edward C. Franklin

Charles L Parsons

Fred E. Wright

Alice C. Fletcher

Arthur Schuster

S. S. Voorhees

Frank Springer

Thomas W. Smillie

Oliver P Hay

Louis T More

W. F. Magie

Wm Mulchey

W. E. Castle

George K Burgess

J. Walter Fewkes

John R. Swanton

Oswald Schreiner

By Ernest Dorsey

Walter Hough

Samuel Henshaw

Arthur A. Noyes

John Johnston

William Whiting

S. F. Acree

W H Holmes

Doctr. Edwin Bidwell Wilson

Frank D. Adams

Jacques Loeb

R B Owens

Geo. S. Huntington M.D.
R A Harper
Claude S. Hudson
Augustus Trowbridge
L. H. Baekeland D. Sc
Henry S. Pritchett
Hon. Francis G. Newlands
Una White
Geo H Ashley
Geo. E. Hale
Mr + Mrs. William B. Hale
Mrs. & Miss W. Harto
W. W. Howell
Frank Morley
F. P. Mall

Charles K. Wead
Robt T. Hill
J H McBride
Dr. W. W. Keen
Arthur Keith
Prof. Thomas Hunt Morgan
H W Henshaw
Vernon Bailey
B. A. Bean
E. O. Hovey
Marion True
H. Bryn, Minister of Norway
Mme Bryn
W. T. Porter
Theo. Gill

THREE CENTURIES
OF
SCIENCE IN AMERICA

An Arno Press Collection

Adams, John Quincy. **Report of the Secretary of State upon Weights and Measures.** 1821.

Archibald, Raymond Clare. **A Semicentennial History of the American Mathematical Society: 1888-1938** *and* **Semicentennial Addresses of the American Mathematical Society.** 2 vols. 1938.

Bond, William Cranch. **History and Description of the Astronomical Observatory of Harvard College** *and* **Results of Astronomical Observations Made at the Observatory of Harvard College.** 1856.

Bowditch, Henry Pickering. **The Life and Writings of Henry Pickering Bowditch.** 2 vols. 1980.

Bridgman, Percy Williams. **The Logic of Modern Physics.** 1927.

Bridgman, Percy Williams. **Philosophical Writings of Percy Williams Bridgman.** 1980.

Bridgman, Percy Williams. **Reflections of a Physicist.** 1955.

Bush, Vannevar. **Science the Endless Frontier.** 1955.

Cajori, Florian. **The Chequered Career of Ferdinand Rudolph Hassler.** 1929.

Cohen, I. Bernard, editor. **The Career of William Beaumont and the Reception of His Discovery.** 1980.

Cohen, I. Bernard, editor. **Benjamin Peirce: "Father of Pure Mathematics" in America.** 1980.

Cohen, I. Bernard, editor. **Aspects of Astronomy in America in the Nineteenth Century.** 1980.

Cohen, I. Bernard, editor. **Cotton Mather and American Science and Medicine: With Studies and Documents Concerning the Introduction of Inoculation or Variolation.** 2 vols. 1980.

Cohen, I. Bernard, editor. **The Life and Scientific Work of Othniel Charles Marsh.** 1980.

Cohen, I. Bernard, editor. The Life and the Scientific and Medical Career of Benjamin Waterhouse: With Some Account of the Introduction of Vaccination in America. 2 vols. 1980.

Cohen, I. Bernard, editor. Research and Technology. 1980.

Cohen, I. Bernard, editor. Thomas Jefferson and the Sciences. 1980.

Cooper, Thomas. Introductory Lecture *and* A Discourse on the Connexion Between Chemistry and Medicine. 2 vols. in one. 1812/1818.

Dalton, John Call. John Call Dalton on Experimental Method. 1980.

Darton, Nelson Horatio. Catalogue and Index of Contributions to North American Geology: 1732-1891. 1896.

Donnan, F[rederick] G[eorge] and Arthur Haas, editors. A Commentary on the Scientific Writings of J. Willard Gibbs *and* Duhem, Pierre. Josiah-Willard Gibbs: A Propos de la Publication de ses Mémoires Scientifiques. 3 vols. in two. 1936/1908.

Dupree, A[nderson] Hunter. Science in the Federal Government: A History of Policies and Activities to 1940. 1957.

Ellicott, Andrew. The Journal of Andrew Ellicott. 1803.

Fulton, John F. Harvey Cushing: A Biography. 1946.

Getman, Frederick H. The Life of Ira Remsen. 1940.

Goode, George Brown. The Smithsonian Institution 1846-1896: The History of its First Half Century. 1897.

Hale, George Ellery. National Academies and the Progress of Research. 1915.

Harding, T. Swann. Two Blades of Grass: A History of Scientific Development in the U.S. Department of Agriculture. 1947.

Hindle, Brooke. David Rittenhouse. 1964.

Hindle, Brooke, editor. The Scientific Writings of David Rittenhouse. 1980.

Holden, Edward S[ingleton]. Memorials of William Cranch Bond, Director of the Harvard College Observatory, 1840-1859, and of his Son, George Phillips Bond, Director of the Harvard College Observatory, 1859-1865. 1897.

Howard, L[eland] O[sslan]. Fighting the Insects: The Story of an Entomologist, Telling the Life and Experiences of the Writer. 1933.

Jaffe, Bernard. Men of Science in America. 1958.

Karpinski, Louis C. Bibliography of Mathematical Works Printed in America through 1850. Reprinted with Supplement and Second Supplement. 1940/1945.

Loomis, Elias. **The Recent Progress of Astronomy: Especially in the United States.** 1851.

Merrill, Elmer D. **Index Rafinesquianus: The Plant Names Published by C.S. Rafinesque with Reductions, and a Consideration of his Methods, Objectives, and Attainments.** 1949.

Millikan, Robert A[ndrews]. **The Autobiography of Robert A. Millikan.** 1950.

Mitchel, O[rmsby] M[acKnight]. **The Planetary and Stellar Worlds: A Popular Exposition of the Great Discoveries and Theories of Modern Astronomy.** 1848.

Organisation for Economic Co-operation and Development. **Reviews of National Science Policy: United States.** 1968.

Packard, Alpheus S. **Lamarck: The Founder of Evolution; His Life and Work.** 1901.

Pupin, Michael. **From Immigrant to Inventor.** 1930.

Rhees, William J. **An Account of the Smithsonian Institution.** 1859.

Rhees, William J. **The Smithsonian Institution: Documents Relative to its History.** 2 vols. 1901.

Rhees, William J. **William J. Rhees on James Smithson.** 2 vols. in one. 1980.

Scott, William Berryman. **Some Memories of a Palaeontologist.** 1939.

Shryock, Richard H. **American Medical Research Past and Present.** 1947.

Shute, Michael, editor. **The Scientific Work of John Winthrop.** 1980.

Silliman, Benjamin. **A Journal of Travels in England, Holland, and Scotland, and of Two Passages over the Atlantic in the Years 1805 and 1806.** 2 vols. 1812.

Silliman, Benjamin. **A Visit to Europe in 1851.** 2 vols. 1856

Silliman, Benjamin, Jr. **First Principles of Chemistry.** 1864.

Smith, David Eugene and Jekuthiel Ginsburg. **A History of Mathematics in America before 1900.** 1934.

Smith, Edgar Fahs. **James Cutbush: An American Chemist.** 1919.

Smith, Edgar Fahs. **James Woodhouse: A Pioneer in Chemistry, 1770-1809.** 1918.

Smith, Edgar Fahs. **The Life of Robert Hare: An American Chemist (1781-1858).** 1917.

Smith, Edgar Fahs. **Priestley in America: 1794-1804.** 1920.

Sopka, Katherine. **Quantum Physics in America: 1920-1935** (Doctoral Dissertation, Harvard University, 1976). 1980.

Steelman, John R[ay]. **Science and Public Policy: A Report to the President.** 1947.

Stewart, Irvin. **Organizing Scientific Research for War: The Administrative History of the Office of Scientifc Research and Development.** 1948.

Stigler, Stephen M., editor. **American Contributions to Mathematical Statistics in the Nineteenth Century.** 2 vols. 1980.

Trowbridge, John. **What is Electricity?** 1899.

True. Alfred. **Alfred True on Agricultural Experimentation and Research.** 1980.

True, F[rederick] W., editor. **The Semi-Centennial Anniversary of the National Academy of Sciences: 1863-1913** *and* **A History of the First Half-Century of the National Academy of Sciences: 1863-1913.** 2 vols. 1913.

Tyndall, John. **Lectures on Light: Delivered in the United States in 1872-73.** 1873.

U.S. House of Representatives. **Annual Report of the Board of Regents of the Smithsonian Institution...A Memorial of George Brown Goode together with a selection of his Papers on Museums and on the History of Science in America.** 1901.

U.S. National Resources Committee. **Research: A National Resource.** 3 vols. in one. 1938-1941.

U.S. Senate. **Testimony Before the Joint Commission to Consider the Present Organizations of the Signal Service, Geological Survey, Coast and Geodetic Survey, and the Hydrographic Office of the Navy Department.** 2 vols. 1866.